The Impact of Climate Change on Ecosystems and Species: Terrestrial Ecosystems

RIVM - RIJKSINSTITUUT VOOR VOLKSGEZONDHEID EN MILIEUHYGIENE

The National Institute of Public Health and Environmental Protection (RIVM) is the key reference institute of the Netherlands Ministry of Housing, Physical Planning and Environment in the area of climate change and other global environmental problems. One of the core activities of the RIVM is to describe the state of the environment and the future environmental quality at the national, European and global levels under various scenarios.

UNITED STATES ENVIRONMENTAL PROTECTION AGENCY

The Climate Change Division in the Office of Policy Analysis at the US Environmental Protection Agency is responsible for projects supporting the inplementation of the Framework Convention on Climate Change and the US Climate Change National Action Plan. Topics covered include the effects of climate change; methods for assessing greenhouse gas sources and sinks; benefiits generated by preventing, or adapting to, climate change; and policies for the reduction of greenhouse gases.

WWF - WORLD WIDE FUND FOR NATURE

WWF - World Wide Fund for Nature is the world's largest private international conservation organisation with 28 Affiliate and Associate National Organisations around the world and over 5.2 million regular supporters. WWF aims to conserve nature and ecological processes by preserving genetic, species and ecosystem diversity; by ensuing that the use of renewable natural resources is sustainable both now and in the longer term; and by promoting actions to reduce pollution and wasteful exploitation and consumption of resources.

THE MARINE AND COASTAL AREAS PROGRAMME

IUCN's Marine and Coastal Areas Programme was established in 1985 to promote activities which demonstrate how conservation and development can reinforce each other in marine and coastal environments; conserve marine and coastal species and ecosystems; enhance awareness of marine and coastal conservation issues and management; and mobilise the global conservation community to work for marine and coastal conservation. The Marine Conservation and Development Reports are designed to provide access to a broad range of policy statements, guidelines, and activity reports relating to marine issues of interest to the conservation and development community.

The Impact of Climate Change on Ecosystems and Species: Terrestrial Ecosystems

Edited by John Pernetta, Rik Leemans, Danny Elder
and Sarah Humphrey
1995

Published by: IUCN, Gland, Switzerland
in collaboration with WWF, US-EPA, SIDA and RIVM

Citation: Pernetta, J.C.; Leemans, R.; Elder, D. & Humphrey, S. (Eds). 1995. *Impacts of Climate Change on Ecosystems and Species: Terrestrial Ecosystems.* IUCN, Gland, Switzerland. viii +99 pp.

ISBN: 2-8317-0171-6

Printed by: SADAG, Bellegarde-Valserine, France

Cover Picture: Map indicating global variation in land vegetation coverage under current climate conditions/RIVM

Layout by: Sarah Humphrey, IUCN

Available from: IUCN Communications Division
Rue Mauverney 28,
1196 Gland, Switzerland

Contents

Preface

This second volume in a series of four concerned with the impacts on climate change on ecosystems and species addresses terrestrial ecosystems. The papers presented result from two meetings, the Symposium on the Impacts of Climate Change on Ecosystems and Species organised in Amersfoort, The Netherlands, in December 1991, and a related workshop convened at the IVth World Parks Congress held in Caracas, Venezuela, in January 1992.

The first three papers in the volume are concerned with tropical systems. Halpin and Smith discuss the possible changes in ecoclimatic zonation in forest areas of Costa Rica under two regional climate change scenarios. The study indicates that shifts in zonation are likely in a range of vegetation types over significant areas of humid tropical forest, including complete loss of some ecoclimatic zones. The management response to an uncertain range of impacts should be to leave options open to the greatest extent possible, and in particular to maintain the natural resilience to change of a healthy forest ecosystem.

In their paper on the miombo woodlands of Zambezian Africa, Desanker and Prentice discuss the application of a gap model to examine the dynamics of the system under climatic change scenarios. The gap model is particularly valuable in that it has relatively low data requirements. The authors conclude that disruption of the tree establishment phase in woodland gaps will have more effect on the system that overall changes in annual climatic parameters.

Williams, Cook, Braithwaite, Anderson and Corbett examine the likely impacts of predicted climate change on the biota of the wet-dry savannas which occupy approximately 30% of the Australian continent. The authors conclude that the predicted change in rainfall, which is the principal environmental parameter which determines savannah function, is small when compared with existing temporal and spatial variation. However, the seasonal variation in rainfall, which at present cannot be predicted, may prove more significant to the savanna flora and fauna. A further unknown factor is the degree to which an increase in CO_2 may favour the woody C_3 plants over the grassy C_4 plants.

The fourth paper by Tchebakova, Monserad, Leemans and Nazimova, describes the use of a regional vegetation model to assess vegetation changes under predicted climatic change in Siberia. The model illustrates a general northwards shift of vegetation zones, with temperate species entering the south of the region, and in mountainous areas, an increase in altitude forspecific vegetation zones. Statistical analysis of the degree of similarity in existing and predicted vegetation maps indicates a wholesale change in vegetation patterns following climatic change.

The final paper addresses a problem generic to all climatic zones undergoing environmental change, namely that of advances in pests which may have a competitive advantage under new or rapidly changing environmental conditions. Pest species - those which are considered to have an adverse effect on the human environment - may be expected to repand to climate change in one of four ways: by spreading into new areas in association with shifts in ecosystems; by occupying niches left vacant by other displaced or debilitated species; by changes in distribution following extreme events such as flooding; and through increased epidemics through vector spread. With climate change, many such changes will be unavoidable and unmanageable, and the author concludes that we need urgently to address the causes of climate change to forestall such events.

The papers in this volume illustrate the range of and uncertainty concerning environmental changes which may be expected to occur as a result of climatic change and increases in atmospheric carbon dioxide concentrations. Although the current techniques and approaches are adapted to fit specific requirements, their prediction of the consequenses of these changes have a more general applicability. The papers also highlight the need by planners and managers for more detailed information about the temporal and spatial nature of such change. In addition to global change, several papers emphasise the impacts of the combination with other human-caused stresses on natural systems. In particular it is this combination that limits the adaptability or resilience of natural systems.

Potential impacts of climate change on forest protection in the humid tropics: a case study of Costa Rica

P.N. Halpin

Global Systems Analysis Program, Department of Environmental Sciences
University of Virginia

C.M. Secrett

International Institute for Environment and Development

Abstract

An assessment of potential changes in ecoclimatic zonation in the forest areas of Costa Rica was conducted using two regional climate change scenarios. This pilot study identified changes in ecoclimatic zones for current biological reserves, natural areas and managed forests in order to evaluate the distribution of protected forest at risk to changes in regional climate. The findings of this analysis indicate that significant areas of humid tropical forest would experience shifts in climatic zonation across a wide range of existing vegetation types under the climatic change scenarios tested. Three findings of this study were that: 1) a significant decrease in the spatial distribution of regions of tropical montane climatic types occurred under the scenarios tested; 2) the changes in ecoclimatic zonation were non-symmetrical along altitudinal gradients; and 3) potential secondary changes in landuse could create additional stresses on forest protection areas. The changes mapped in this pilot study indicate that a potential loss in landscape scale patterns of habitat diversity may occur in the neotropical regions under projected climate change scenarios. This type of potential ecoclimatic change coupled with ongoing changes in landuse may act to significantly increase stress on the management of currently protected habitats in the region.

Introduction

Analyses of potential changes in vegetation patterns in response to possible climate change have generally been conducted at global or continental scales with relatively low spatial resolutions (Emmanuel *et al.*, 1985, Leemans & Cramer 1990; Smith *et al.*, 1990; Prentice *et al.*, 1989; Prentice *et al.*, 1992; Leemans & Halpin, 1992). While these modelling efforts have generally succeeded in identifying gross trends in potential vegetation change, they generally fail to depict potential impacts of climatic change at spatial scales appropriate to the needs of natural resource planning and conservation management. Adaptive policy responses to potential impacts of climate change on different sectors of forest management can only be analysed at a regional scale with a specific focus on the unique forest issues of each region.

A first step toward the development of more detailed assessments of climate change and tropical forest systems was conducted for the neotropical forest area of Costa Rica. This case study was undertaken as an initial pilot project to develop techniques for the analysis of vegetation change at regional and subregional scales. The case study goes beyond provisional mapping of changes in potential vegetation classes by analysing the spatial changes in ecoclimatic zones in relation to existing natural forest areas, the level of protection of the forest areas and the land use classification of the sites (Halpin *et al.,* 1991).

This pilot project focuses on forest production and nature reserve impact assessment. The Costa Rica pilot project was conducted with formal collaboration and generous assistance of numerous in-country participants including members of the Tropical Science Center (TSC), the National Meteorological Institute (IMN), the National Parks Service (SPN) the Forest Service (DGF), Fundacion Neotropical (FN), the Organisation for Tropical Studies (OTS), and the regional office of the World Conservation Union (IUCN-ORMA). Detailed workshops with all participants were conducted in Costa Rica. Central American forest overviews and potential tropical forest policy responses presented in this review rely heavily on collaborative work of the International Institute for Environment and Development (Secrett 1991, 1992).

Site Selection

Costa Rica was selected as an ideal tropical case study site because it exhibits a relatively complete microcosm of tropical climate regimes, ecosystem types, land use pressures, and natural resource conditions representative of a wide range of humid tropical regions in the developing world. Many general findings of this case study can be reasonably extrapolated to similar climatic regions throughout the humid tropics. Costa Rica offers an extensive range of distinct ecoclimatic zones and patterns of landscape diversity due to its exceptional topographic relief (0 - 3,800 m) and position on the thermal equator (8° to 10° North latitude) (Holdridge, 1967; Gomez & Herrera, 1986; Hartshorn, 1983; Walter, 1985). The forest systems of Costa Rica contain noteworthy examples of a wide range of forest types from the relatively rare tropical dry forest sites of Guanacaste in the northern Pacific region, to lowland and montane tropical wet forests in the Atlantic coastal regions. This diversity of forest systems sets Costa Rica apart as a living laboratory for climate-vegetation analysis.

Because Costa Rica's expansive nature reserve system represents sites spanning numerous climatic and orographic gradients, the country presents a unique situation for the analysis of potential climatic impacts on ecosystem and biodiversity protection in the humid tropics (Halpin, 1992). Costa Rica contains some of the finest examples of tropical montane nature reserves and some of the few existing protected area corridor systems in the tropics (Mackintosh *et al.*, 1989).

Finally, Costa Rica is one of the few locations in the tropics which has a well-developed and long-standing history of climate-vegetation analysis and field validated ecoclimatic mapping relating climate to natural vegetation, productive forestry and agricultural land use (Holdridge 1967; Holdridge *et al.*, 1971; Sawyer & Lindsey, 1971; Gomez & Herrera, 1986). Ecoclimatic mapping for use in forestry and biodiversity management, land capability analysis and natural resource accounting serves as a vital component in existing natural resource assessment techniques in the country (Holdridge & Tosi, 1972; Hartshorn *et al.*, 1982; Tosi *et al.*, 1988; Sader & Joyce, 1988; TSC/WRI, 1991; Cruz & Repetto, 1992). Because land capability and natural resource accounting techniques used in Costa Rica already incorporate climatic features into the land planning system, analysis of climatic changes on natural resources is more readily incorporated into long range national assessments. This basis of past analysis and ground verification of ecoclimatic classifications coupled with ongoing forest land use debates using climatically derived land capability accounting systems makes Costa Rica an exceptionally rich site for preliminary climate change assessment.

Forest Background

With average precipitation exceeding 3,000 mm and annual temperatures ranging from 27.8°C at sea level on the Pacific Coast to 4.5°C at the highest mountain summits, Costa Rica's original vegetation represented exceptional examples of humid broadleaf tropical forest types. Of the 12 general ecoclimatic zones found in the country, only the high altitude, sub-alpine paramo regions (0.2% of the country) are naturally without full forest cover (Hartshorn *et al.*, 1982). Over 4,000 species of forest trees and shrubs are estimated to exist in the region and an exceptionally wide variety of endemic flora and fauna species are found in the forested areas (eg. 848 species of birds, 216 species of reptiles, 205 species of mammals and 160 species of amphibians) (Janzen, 1983; Hartshorn, 1983).

The extent of the remaining forest area in the country has been severely diminished in the last century through land clearing for agricultural, pasture and plantation use. (Sader & Joyce, 1988; Tosi *et al.*, 1992). A recent World Bank/UNDP study (1990) estimates that current deforestation is due mainly to agricultural expansion (70%), commercial logging (20%) and fuelwood collection (5%). Historic mapping and recent satellite remote sensing of forest cover have been employed to document the decline of the forest cover in recent years. Figure 1 depicts the loss of primary forest cover in Costa Rica from 1940 to 1987 (Fundacion Neotropica, 1988).

Of special interest to climatic change analysis is the fact that past trends in deforestation have been directly associated with the ecoclimatic characteristics of the forest sites, with life-zone climate types suitable for agricultural uses being preferentially cleared (Sader & Joyce, 1987; TSC/WRI, 1991). The highest rates of recent deforestation have occurred in the tropical dry forest and pre-montane moist forest climate types while the lowest rates of deforestation have occurred in the pre-montane rain, lower-montane rain and montane rain forest sites (Sader & Joyce, 1987). High rainfall, low potential evapotranspiration to precipitation ratios and inaccessible terrain of the montane climate zones have acted together to discourage widespread deforestation in those regions. A recent natural resource accounting assessment conducted by

Figure 1.
Deforestation in Costa Rica 1940 - 1987.

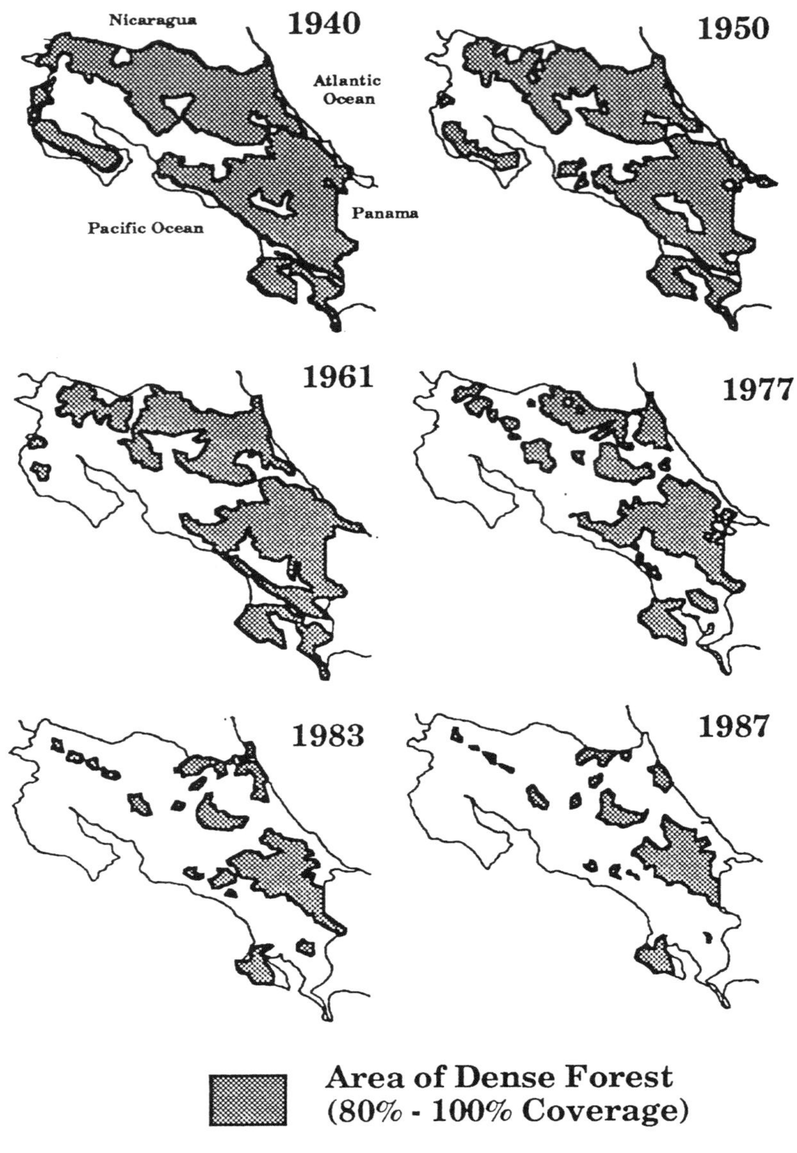

the Tropical Science Center and the World Resources Institute (1991) concludes that the largest areas of cumulative deforestation occurred in pre-montane wet (285,785 km^2) and low-land tropical wet forest (212,945 km^2) sites. These areas are generally considered to contain the highest potential for biodiversity in the region (TSC/WRI, 1991). General deforestation rates have been estimated at approximately 3.9% of remaining forest area per year (WB/UNDP, 1990). Rates have dropped from approximately 500 km^2 per year (1950-1984) to approximately 318 km^2 per year due to changes in land use policies and increased inaccessibility of remaining forest areas (Secrett, 1991; TSC/WRI, 1991, PAFT-CR, 1990). It is estimated that by the year 2060, all remaining primary forest coverage will be contained in the reserve system and most forests outside of strict reserves will be made up of managed secondary growth stands (Tosi *et al.*, 1992).

Estimates of current forest extent and land capability ratings differ between sources, but a recent remote sensing study estimates that the land surface under forest coverage has been reduced from 53% in 1950 to 29% in 1987 (DGF, 1988). Of this remaining area only 5.3% is estimated to be have value for commercial timber production while the remainder is classified as suitable only for protection or forest buffer regions. The majority (70%) of closed canopy forest areas remaining in the country are now contained within established reserve areas while the remainder are fragmented into patches of less than 10 km^2 (WB/UNDP, 1990).

The expected economic value of forest resources lost due to deforestation have been recently calculated for Costa Rica as $159.3 million ($66 per capita) lost by 1984. Also for each 1000 colones of production derived from pasture uses on once forested lands, 136.5 colones worth of production is lost due to soil erosion (TSC/WRI, 1991). Between 1988-89, the estimated depreciation of Costa Rica's forest assets was 36% more than all external debt service payments by the public sector (TSC/WRI, 1991). These natural resource losses have not been costed in the national accounting system.

Types of forest land uses in Costa Rica have been broadly defined for this study as intensive forestry, extensive forestry or protection forestry management categories (Tosi *et al.*, 1988; TSC, 1992). Intensive forestry uses are defined as highly managed production forestry and short rotation plantation forestry management methods. Extensive forestry uses are defined as natural forest management and secondary forest regrowth uses exhibiting limited management intrusions and harvesting disruption. Protection forest uses are classified in a hierarchy of strict biological reserves, wildland areas, forest reserves and indigenous indian reserves.

The potential effects of climate change are considered for four general forest management sectors in this pilot project. These are the potential effects of regional climatic change on 1) sustainable forest production; 2) natural area protection; 3) fuelwood production; and 4) forest watershed protection. A brief description of the current status of these forest sectors and general susceptibility of these sectors to climatic disruption are discussed below.

Forest Production

Total forest production capacity in Costa Rica has remained fairly constant in recent years with 1988 output at 3,961,000 m^3 with 35% of this used for industrial purposes (sawnwood, panels, pulp and paper) and 65% used for local purposes (fuelwood, construction posts, fencing) (IDB, 1987). Industrial forest production is concentrated on the production of sawnwood (70% of total volume), panels and veneers (17%) and paper products (13%). Virtually all sawnwood products are consumed domestically (99%) while panel products are exported. Over 90% of paper products used in the country are imported leading to the recognition that Costa Rica is a net importer of wood products with a $39.7 million trade deficit in this sector in 1985 (IDB, 1987).

While Costa Rica contains approximately 2,000 species of trees, only 70 are commercially used in timber production and of these only a small percentage comprise the bulk of the trade (Secrett, 1991; Hartshorn *et al.*, 1982; WB/UNDP, 1990). This species selection coupled with high transportation costs, aging equipment, unsustainable forest extraction methods and a lack of management incentives has led to a generally inefficient forestry practices with approximately

46% of commercial timber volume lost during extraction (Flores, 1985; TSC/WRI, 1991). Increasing forestry efficiency and long term sustainability of forest uses are critical issues in the production forest sector at present.

Under changing climatic conditions, the ability to meet forest product demands, increase forest use efficiency and change forestry practices to more sustainable methods may be made more difficult by increased environmental and infrastructure stresses.

Natural Area Protection

The loss of the majority of forested area due to land clearing has dramatically increased the need for the immediate protection of forest areas through classification of remaining forest lands into a variety of protected management classes. The current national protected area strategy involves a mix of various types of protected areas in eight general conservation (mega-park) regions which are intended to provide both sustainable production of natural resources and the protection of forest biodiversity (PAFT-CR, 1990). The regional grouping of various reserve areas in national planning is intended to promote the future development of multiple use forest resource management schemes and contiguous buffer zones around sensitive biological reserve core areas. Figure 2 depicts the locations of national parks and biological reserves, forest reserve areas and indigenous Indian reserves within the eight general management regions.

While Costa Rica's forest protected area system is considered to be one of the best in the world, problems in site management and enforcement of existing protection mandates exist (PAFT-CR, 1990). In general, climatic change could significantly exacerbate existing management issues by: 1) directly altering the ecological conditions and species composition of reserve areas; 2) by lowering the natural resilience of ecosystems to disturbance; 3) by increasing external land-use pressures on reserve areas; and 4) by lowering the national capability to support protected area infrastructure and enforcement. These problems would be common to all countries in the region.

Fuelwood/Energy Use

In 1987, 68% of Costa Rica's energy uses were generated within the country. Of this amount, 44% was attributed to fuelwood and charcoal use, 14% to hydroelectric power and 10% to crop residues burning (Calvo, 1990). Of the fuelwood, 80% was consumed by households and 20% by industries.

While domestic demand for fuelwoods is expected to be met in the future for the country as a whole, skewed geographic concentrations of population and fuelwood demand create problems in local production and distribution of this forest resource. Regional impacts of climatic change on forest production and general land use patterns could act to significantly increase existing fuelwood resource problems. Areas such as the Central Valley which supports a high population, high industrial fuelwood use and low forest area would be especially sensitive to direct climatic changes and secondary land use shifts on sustained fuelwood provision. The relatively small size of Costa Rica, however, lessens the impact of transportation of woodfuel resources within the country.

Figure 2.
Forest conservation regions of Costa Rica.

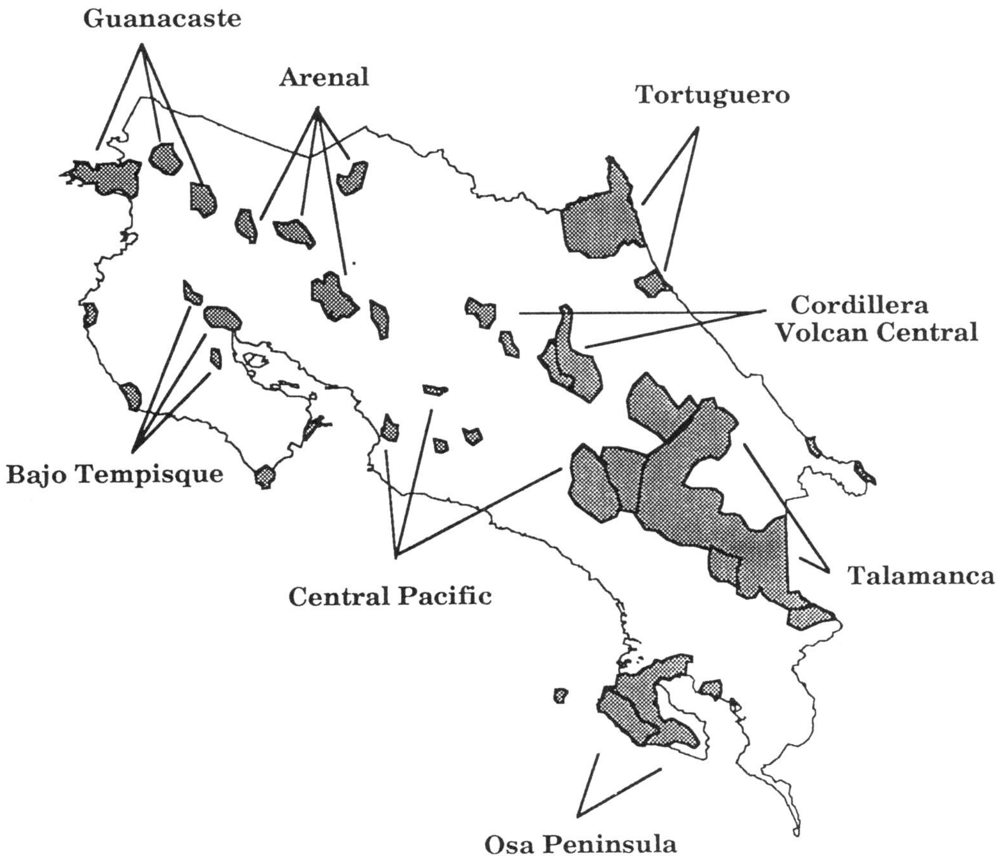

Waterbasin Protection

Forest vegetation serves critical roles in the provision of sustained water resources and the protection of watersheds from bank erosion, stream siltation, soil loss and biodiversity loss. Costa Rica has been divided into 38 distinct watershed areas and the percent of remaining forest cover and area of each waterbasin under protection has been recently assessed (FN/CI, 1988). The Caribbean watersheds contain approximately 65% natural vegetation coverage with 17.3% in protected management. The Pacific waterbasins retain approximately 33% natural vegetation coverage, with only 2.7% in protected areas. The northern and central waterbasins of the Pacific slopes are under the most intensive land use and contain the least natural vegetation with 12 waterbasins having no lands under protection. These waterbasins are potentially more at risk to significant erosion losses.

Changes in forest structure and function due to regional climatic changes could alter the current evapotranspiration and rainfall-runoff regimes of critical waterbasins in the region, potentially exacerbating soil erosion and water management systems in critical waterbasins.

Approach and Methods

As with many countries in the humid tropics, Costa Rica has faced rapid declines in the extent of natural forest area and an increase in land use pressures on the remaining forest estate in recent years. The potential impacts of climate change on forest production and protection in the humid tropics therefore must be interpreted within the context of continuing changes in forest extent and land use capability (Halpin *et al.*, 1991). This pilot project assesses potential climate change impacts on existing forest management areas, nature reserves and unmanaged natural areas and also identifies potential changes in land use capability for the country as a whole. Changes in land use capability can be used to infer both direct impacts of climate change by suggesting alteration of the productivity of a site under the current land use regime, but also offers useful insights into potential secondary changes in preferential land use which may be brought about by changing climatic conditions.

Direct impacts of climatic change on forests are interpreted as changes in expected productive capability and composition of existing forest areas. Direct climatic impacts are assumed to be produced by increases in physiological heat stress, changes in seasonal water balance and changes in the frequency and magnitude of climatic disturbances. Changes in heat stress and water balance related to global GCM scenarios are assessed directly in this study. While reliable assessments concerning changes in regional disturbance regimes for climate change conditions are not available for Costa Rica, recent studies for the Caribbean region predict increases in both hurricane frequency and magnitude, and subsequent forest damage, due to increased sea surface temperatures (Maul, 1989; Granger, 1989; O'Brien *et al.*, 1992).

Direct climatic impacts on forests are generally defined in terms of changes in mean annual growth increments (MAI) for production forest species, potential ecological stresses on protected forest areas, changes in seasonality of climatic factors, and changes in hydrologic functions for forested catchments.

Indirect or secondary impacts of climate change, on the other hand, can be defined as changes in potential land use as a response to changing climate. For example, if the climate regime of lands currently used for cattle pasture are expected to become better suited for secondary forest regrowth, economic pressures to convert the area back to forestry uses may also increase. On the other hand, a potential negative response in terms of forest use could occur if agricultural lands adjacent to forest stands lose per area productivity (land capability) due to climatic change, forcing an expansion of agricultural zones into forest areas in order to offset potential agricultural losses.

Both direct and indirect responses to climate change must be considered in any analysis of potential climate change impacts in tropical regions. In this analysis, spatial areas of existing ecoclimatic zones were modified to represent the outcomes of potential climatic changes on a regional scale. Direct effects on production forests and nature reserves were assessed through geographic analysis and secondary impacts in terms of potential changes in land capability were defined through expert interpretation on a sub-regional basis by in-country collaborators (Halpin *et al.*, 1991; Tosi *et al.*, 1992). Direct impacts of the scenarios tested were assessed using the ecoclimatic life-zone model of Holdridge (1949, 1967).

Holdridge Life Zone Classification

Features of the Holdridge Classification are summarised in figure 3. Ecoclimatic life- zones are depicted by a series of hexagons formed by intersecting intervals of climate variables on logarithmic axes in a triangular coordinate system. Two variables, average biotemperature and average annual precipitation, determine the classification. Average biotemperature is the average temperature over a year with the unit temperature values (ie. daily, weekly or monthly) that are used in computing the average set to 0°C if they are less than or equal to 0°C. In the Holdridge Diagram (Figure 3), identical axes for average annual precipitation form two sides of an equilateral triangle. A logarithmic axis for the potential evapotranspiration (PET) ratio (effective humidity) forms the third side of the triangle, and an axis for mean annual biotemperature is oriented perpendicular to the base. By marking equal intervals on these logarithmic axes, hexagons are formed that designate the Holdridge Life Zones. Each life zone is named to indicate a vegetation association and represents the climate conditions or ecoclimatic zone associated with that ecosystem or vegetation type. The terms life zone and ecoclimatic zone will therefore be used interchangeably throughout the discussion of results.

Potential evapotranspiration is the amount of water that would be released to the atmosphere under conditions of sufficient but not excessive available water throughout the growing season. The potential evapotranspiration ratio is the quotient of PET and average annual precipitation. Holdridge (1959) assumes, on the basis of studies of several ecosystems, that PET is proportional to biotemperature. The PET ratio in the Holdridge Classification is therefore dependent on the two primary variables, annual precipitation and biotemperature.

Figure 3.
The Holdridge Life-Zone Model (from Holdridge 1949, 1967).

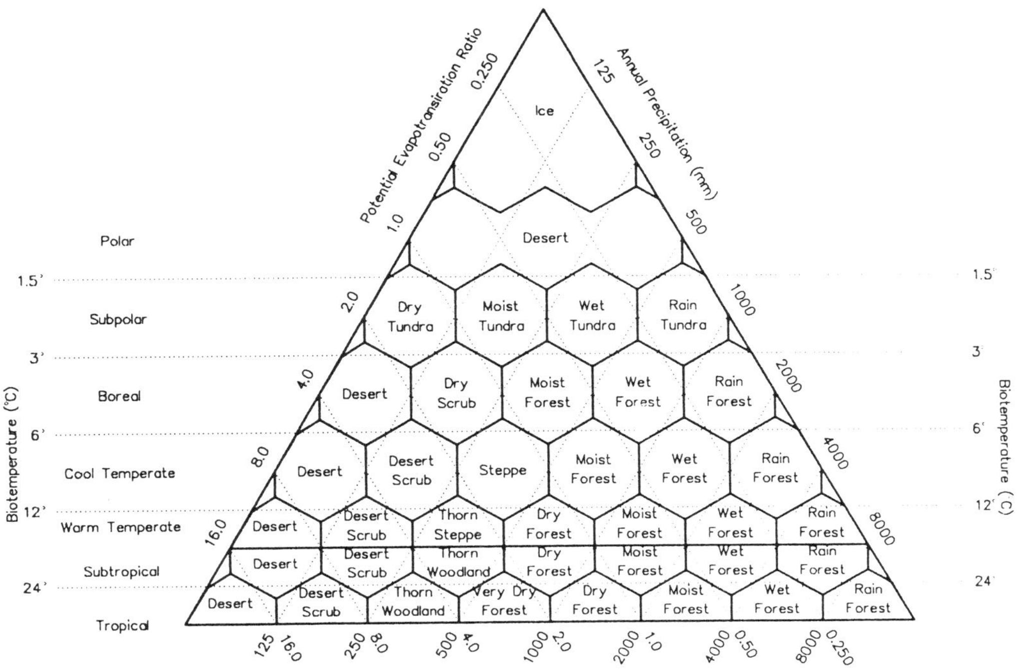

One additional division in the Holdridge System is based on the occurrence of killing frost. This division is along a critical temperature line that divides hexagons between 12°C and 24°C into warm temperate and subtropical zones. This line is adjusted to reflect regional conditions. The complete Holdridge Classification at this level includes 37 life zones.

The Holdridge model, like all ecoclimatic models, is essentially a correlation between climate and vegetation patterns. Although it provides a convenient description of the association between features of the vegetation and regional climate, the model implicitly assumes that vegetation patterns are effectively at equilibrium with current climate patterns. Although this may be an essentially valid assumption under current conditions, it presents a problem in the application of this class of models to the climate change scenarios investigated in this study. In many cases, the timescale associated with the changes in climate (ie. 70-100 yrs) is much faster than the processes required for vegetation to respond (eg. long-distance migration of species) to these new patterns. Therefore, we restrict our use of this model to investigate the implications of the climate change scenarios on the distribution of ecoclimatic zones associated with current vegetation or ecosystem types. *This method provides a means of evaluating the potential for a geographical area to sustain a particular vegetation type, not whether a vegetation type will occur there within the timeframe associated with the climate change scenario. A major change in ecoclimatic classification for a site may be interpreted only as an indication of potential ecological stress, but not a direct transfer of ecosystem type.* Despite this restrictive use, the methodology provides a valuable tool for assessing the implications of climate change on the current distribution of vegetation and the potential for change in vegetation patterns.

Two specific approaches to mapping the potential impacts of climate change on Costa Rica were undertaken. The first, based on a spatial analysis of climate, topography and land use data was conducted by the Global Systems Analysis Program at the University of Virginia (Halpin *et al.*, 1991; Halpin, 1992); and a second approach relying on a combination of climate data and expert interpretation of existing vegetation site features was conducted by the Tropical Science Center, San Jose, Costa Rica (Tosi *et al.*, 1992). Initial mapping was conducted at the University of Virginia at a map scale of 1:500,000 (IG/CR, 1987) using climate data from a national network of meteorological stations (IMN, 1985; IMN Climate Data: Campos, pers. comm., 1990; Portig, 1969). These climate data were combined with soils and topography information to develop a provisional life zone map for the region. Climate change values were derived and used to modify the spatial composition and extent of potential forest cover types using a high resolution version of the techniques outlined in the global analysis section. Digital mapping of the current distribution of natural vegetation coverage, forest reserves and protected areas were then used to determine the spatial extent of ecoclimatic changes within different land management classes (Halpin *et al.*, 1991; Halpin, 1992).

A second level of analysis was conducted at a 1:200,000 map scale by project collaborators at the Tropical Science Center (San Jose, Costa Rica) through a manual interpretation of micro-climate and site features using a newly revised version of the ecological map of Costa Rica and underlying soils, topography and land capability databases (Tosi, 1969, Tosi *et al.*, 1992). This manual interpretation of ecoclimatic zone changes was then entered into a geographic information system for analysis. The existing (1:200,000) life zone map for Costa Rica, which was developed using a combination of climate data and field observation of present vegetation patterns, provided a significantly higher resolution representation of long-term climate patterns than interpolated climate station data for this extremely mountainous region (Holdridge *et al.*, 1971). The ecoclimatic map was used as a surrogate high resolution climate base map in this analysis.

Regional Climate Scenarios

The regional results of four GCM scenarios tested fell into two distinct groups for the Costa Rican area. Both the GFDL (Manabe & Wetherald, 1987) and the OSU (Schelsinger & Zhao, 1988) models produced regional estimates of approximately a 2.5°C increase in temperature and a 10% increase in precipitation. The UKMO (Mitchel, 1983) and GISS (Hansen *et al.*, 1988) models on the other hand, both produced regional change values of approximately 3.6°C and 10% increase in precipitation.

These paired results defined two working climate change scenarios which fell within the moderate and upper reaches of the regional IPCC based transient estimates for the time period of interest (Kelly, 1991). The two working scenarios were then used to modify the base climate map in a spatially explicit grid format. This process was conducted through a formula driven approach at the University of Virginia and was replicated manually using expert interpretation of site features, topography and soils at the Tropical Science Center (Halpin *et al.*, 1991, Tosi *et al.*, 1992).

Results

The potential impact of the moderate change scenario (+2.5°C and 10% precipitation increase) and the more extreme change scenario (+3.6°C and 10% precipitation increase) on the country-wide distribution of ecoclimatic (ie. life) zones is depicted in Figure 4a and 4b. Ingeneral, the two regional scenarios produced significant changes in the spatial arrangement and regional composition of ecoclimatic zones. The 2.5°C scenario produced a 38% change in ecoclimatic zones while the 3.6°C scenario produced a 47% change. When this analysis was replicated at a higher spatial resolution by the Tropical Science Center, 43% of the land area changed ecoclimatic zones under the 2.5°C scenario and 60% under the 3.6°C scenario (Tosi *et al.*, 1992).

The increase in area found to experience changes in ecoclimatic type with the higher resolution assessment is consistent with expectations that increases in spatial resolution increase the differentiation of distinct ecoclimatic zones, therefore, identifying more areas of potential change. The types of ecoclimatic change identified in the mapping of life zone shifts indicate three generalised trends under both scenarios. First, changes in the areal arrangement of ecoclimatic zones suggest a strong trend toward the displacement of montane and subalpine zones by warmer pre-montane and lower-montane climate types. Second, changes in the calculated annual biotemperature indicates an increase in the period and magnitude of heat stress for many low elevation areas. Third, the entire climatic gradient of the region is seen to experience a change toward more tropical climatic types across all elevations.

Figure 4a.
Area of change in ecoclimatic zone under two precipitation scenarios (Adapted from Tosi , 1992)

(a) a +2.5°C / +10% **(b) a +3.6°C / +10%**

This potential loss of unique ecoclimatic zones is due to higher temperatures reducing the number of distinct zones found along altitudinal gradients. Country-wide assessment of the changes in ecoclimatic zones under the two scenarios represent similar directional trends in terms of the types of expected change, but represent different thresholds of transition and differences in area of change. Under the 2.5°C scenario, the principle changes occur in the warm-dry and cool-wet climate zones, with tropical dry forest, montane rain forest and sub-alpine paramo climatic zones being the most affected areas. The sub-alpine paramo climate zone disappears completely at all scales of analysis under this scenario (Halpin *et al.*, 1991; Tosi *et al.*, 1992).

Under the 3.6°C scenario, lowland tropical moist and wet forest climate zones decrease in spatial area while large areas of increase are expected for premontane forest climate types due to changes in calculated biotemperature for the region. Calculation of expected biotemperatures for the region demonstrate that both scenarios would result in significant alteration of annual growth periods for lowland forest areas (0-900 m) with an expected increase in heat stress moving up the altitudinal gradient. The expected effect of heat stress on vegetation within these zones is a general reduction in annual productivity in the short term and the potential for changes in vegetation structure, biomass and species composition over a longer time period (Holdridge *et al.*, 1971; Tosi *et al.*, 1992).

Sub-Regional Analysis

Five general climatic sub-regions were used in the analysis to calculate specific climate change impacts. Figure 5 depicts these geographically and climatically distinct regions. A summary of the general sub-regional impacts defined by the Tropical Science Center (Tosi *et al.*, 1992) are presented below.

Central Valley

In the Central Valley (800 - 1,200 m) both the 2.5°C and 3.6°C scenarios would produce heat stress effects only on vegetation in the lower elevation extremes, however ecoclimatic zones would be expected to change over much of the region. Changes from pre-montane climate types to tropical basal zones predominate in the region. Vegetation requiring cooler temperatures, such as arabica coffee and ornamental flowers grown for export may be forced to higher elevation zones. This potential change in land use could act to displace dairy and truck crops in the lower montane regions. Attempts to offset potential agricultural losses by encroachment into the higher montane wet forest sites could act to endanger much of the last remaining high elevation forest areas in this region which serve to maintain the nation's most reliable sources of hydroelectric power.

Because this region is the most heavily populated and urbanised area of the country, limited opportunities exist for redistributing crop and forest land area under climate change scenarios. Anticipatory development policies may be best directed toward retaining the highest level of non-urbanised areas as possible into the future in this region.

Pacific South

In the Pacific South region, forest areas would be affected by an increase in heat stress conditions expressed as the difference between ambient air temperatures and calculated biotemperatures (t/tbio) for the region. The changes in biotemperature and increased rainfall would move climate types in this region from tropical pre-montane forest to shorter growing period montane forest climates. Actual evapotranspiration is expected to decrease, the number of excessively wet months is expected to increase and run-off is expected to significantly increase in this region. This condition is expected to lower per hectare forest production for this area. Forestry would continue to be the most suitable land use under these climates, but production would be somewhat diminished.

Atlantic

In the Atlantic Sub-Region, heat stress and biotemperature changes from the current Caribbean climate patterns are not expected to be severe. Forest runoff is expected to increase significantly under both scenarios due to increased rainfall, but the seasonal relationship of wet months is expected to be unchanged. Excessive water in the lowland areas of this region is expected to remain a current hindrance to agricultural land uses. Forest production is expected to remain the most viable land use for this region under both scenarios.

Figure 5.
General climatic regions of Costa Rica.

Pacific North

The dry Pacific North Sub-region is expected to be negatively impacted by changes in heat stress and soil moisture deficit under both scenarios. Areas below 500 m elevation are expected to become a belt of extreme heat stress for cultivated plants and cattle grazing land uses. Native forests in this region have already been reduced to small fragments contained in nature reserves. Relics of these dry forests should be considered as germplasm for reintroduction of species into reclamation areas which may have similar climates in the future.

Northern Subslope

The Northern Subslope represents the most continental climate zone represented by tropical moist forest climate types. While a large percent of the area is now under agricultural use, areas of poor soil, insufficient drainage and less favorable topography are left for forest uses. Under the 2.5°C scenario, increases in heat stress and forest run-off are expected to decrease the suitability of much of the area for agricultural pursuits and increase the land use suitability for forest production use. Under the 3.6°C scenario, heat stress, increasing run-off and an increase in wet months from 5 to 7 months per year produce conditions expected to significantly reduce forest productivity and hamper forest operations, especially commercial logging in the region.

Geographic Impacts on Forest Reserves and Protected Areas

In order to draw more direct conclusions concerning the risks posed by climate change for natural vegetation areas, production forestry sites and existing nature reserves, potential changes in ecoclimatic zones must be assessed in terms of actual land-use distributions. In this study, current areas of natural vegetation coverage, production forest sites and protected areas were assessed for changes in ecoclimatic zonation. Mapping of non-agricultural areas containing 60% to 100% natural vegetation coverage were used to define the potential spatial impacts of changes in ecoclimatic zones on remaining forested vegetation within the country. Figure 6 depicts levels of existing natural cover for Costa Rica. Under the 2.5°C scenario, 31.8% of the remaining natural vegetation areas experienced changing ecoclimatic conditions while the 3.6°C scenario produced 48.2% change in ecoclimatic zones. These non-agricultural areas represent the most likely areas for potential forest management land-use adaptation, such as the development of forest buffer zones, corridor systems and new forest reserve areas under changing climatic conditions.

Production forest areas and forest reserves were found to exhibit 19.5% change in ecoclimatic zones under the 2.5°C scenario and a 44.5% under the 3.6°C scenario. In a more general assessment of production forest potential, the Tropical Science Center report states that major changes in the tropical dry, sub-alpine and montane zones, while dramatic in terms of conservation areas, will have only minor effects on forest production due to a lack of forestry land uses in those zones at present. However, general increases in the area of tropical moist forest climate types is expected to raise the potential opportunity for forest growth in areas now managed for pasture and agricultural uses.

Figure 6.
Areas of current natural vegetation cover.

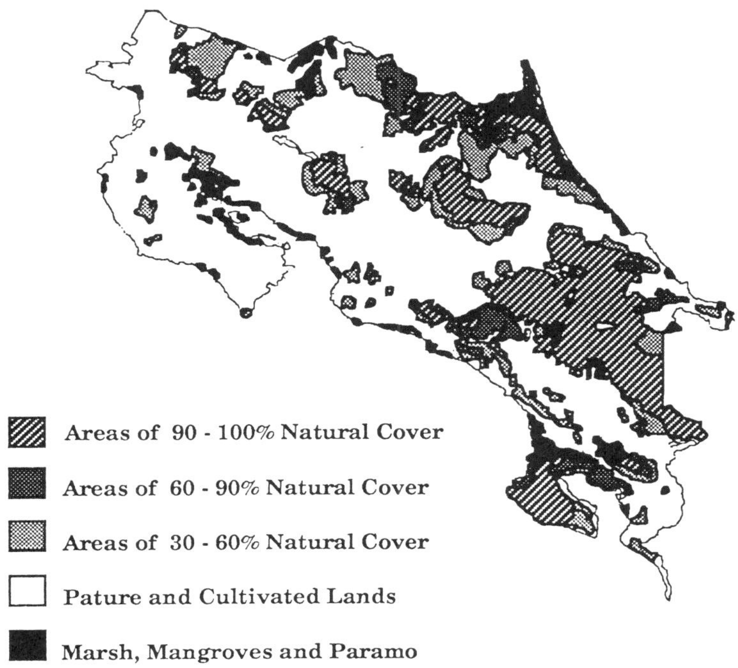

Figure 7.
General altitudinal changes in ecoclimatic zonation along an East-West transect for South-Central Costa Rica.

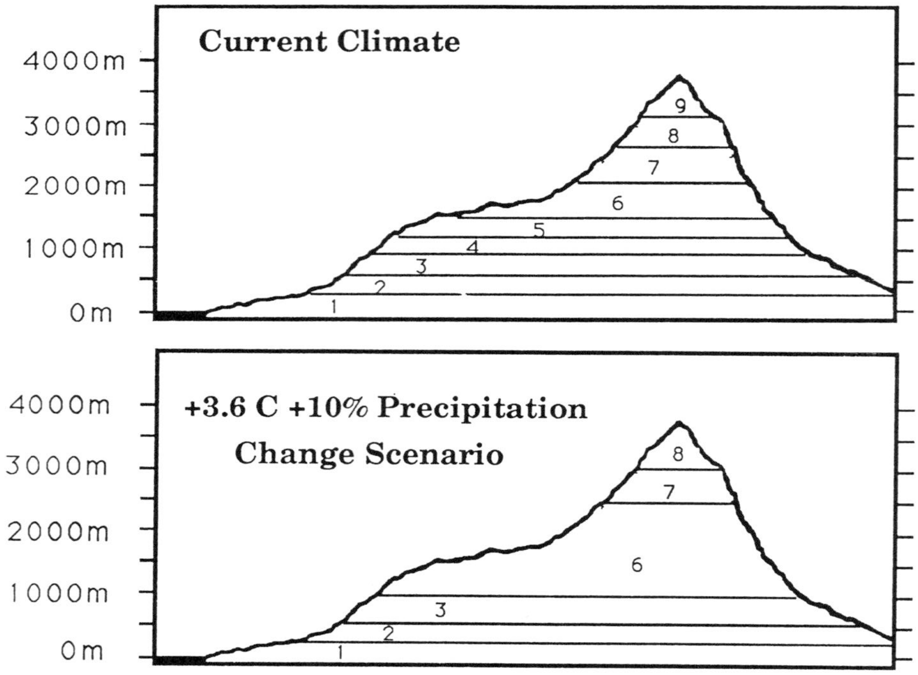

1 Tropical Wet Forest: Warm Transition
2 Tropical Moist Forest
3 Tropical Wet Forest
4 Tropical Wet forest: Cool Transition
5 Premontane Wet Forest: Warm Transition
6 Premontane Rain Forest
7 Lower Montane Rain Forest
8 Montane Rain Forest
9 Sub-Alpine Rain Paramo

The nature reserve system of Costa Rica spans a wide range of topographic and climatic habitat types. For the purpose of this study, digital maps of 33 National Parks and biological reserves were analysed with respect to potential changes in ecoclimatic zones. Under the 2.5°C scenario 27% of the area contained within the reserve system experienced a change in ecoclimatic zone, as compared with 53% for the 3.6°C scenario.

The changes in ecoclimatic zones expected to occur within the major reserve systems represent areas which experience subtle differences in temperature and moisture as well as areas which experience dramatic changes in the potential composition of ecoclimatically defined habitats. For example, the la Amistad reserve complex, located along the mountainous region of southern Costa Rica, experiences both a dramatic change in types and the number of unique climate zones represented in the reserve boundaries. Under the 3.6°C change scenario a significant increase in the ecoclimatic zone associated with present pre-montane forests moves up the slopes displacing unique montane and sub-alpine climate types and transition zones. This displacement of cool montane climatic zones reduces the number of unique ecoclimatic zones within the reserve area from nine to six life zones with pre-montane and lower montaneclimate types accounting for approximately 85% of the area (Figure 7).

This change demonstrates an expected loss of the climate types now associated with the dwarf cloud forest and sub-alpine vegetation communities currently occupying the higher altitude areas of the reserve system and the potential for significant dislocation of forest vegetation and associated faunal species within reserve areas.

While analysis of changes in the ecoclimatic zones associated with present vegetation distributions offers a reasonable method for assessing general trends in the climatic features which effect floral and faunal habitats, this approach does not attempt to identify the particular responses of individual populations and species groups to specific climate changes. Individual species will obviously respond to climatic change through independent processes of acclimatisation, migration or adaptation (Peters & Darling, 1985; Hunter *et al.*, 1988; Graham, 1988; Davis, 1989). In this analysis, a change from an ecoclimatic zone presently associated with a specific vegetation type to another is interpreted as a sign of potential ecological stress, not an explicit prediction of the future vegetation cover. The time frame of the change predicted by the GCMs for the regional change scenarios (eg. 2060-2080) does not allow for an assessment of potential vegetation in terms of expected equilibria with future climate (Halpin *et al.*, 1991). The expected response times of canopy dominant tropical forest trees is in general far longer than the expected rate of change in climatic zonation.

While the ability of individual species to tolerate rapid changes in climate are not generally known for the region, it can be assumed that these rapid changes in ecoclimate would have a significant effect on individual species in terms of net primary production, regeneration, competitive ability and dispersal. These potential changes in individual species responses to changes in climate zone could contribute to local extinctions of particular forest species, the creation of novel species associations and significant changes in the abundance of highly specific habitat types often associated with much of the fauna of the neotropical region (Halpin, 1992).

General Implications

General implications assessed by the Global Systems Analysis Program and the Tropical Science Center are summarised below:

- The economies of the forestry sector (both timber and fuelwood) and of the national system of protected areas will be particularly sensitive to any future climate changes due to their total dependencies on the natural environment.

- Global climatic change would significantly modify ecoclimatic zones in the region. Land-use and forestry capability classifications would be extensively affected by the changes assumed in the scenarios with significant heat stress occurring in low elevation forest areas.

- Three life zone types in the hot-dry and cold-wet extremes disappear from Costa Rica under the scenarios evaluated, indicating the possibility of significant impacts on the maintenance of habitats and species diversity in those regions.

- Loss of distinct ecoclimatic zones along altitudinal gradients in the nature reserves ystem demonstrate a potential for a reduction in the landscape level diversity of critical habitats maintained within the reserve system.

- Agricultural production and cattle ranching land uses would likely be forced to higher elevations due to potential heat stress. This could likely displace existing fragile forest areas remaining at these elevations.

- Future wood production is expected to decline on a per hectare basis due to the direct impacts of increased heat stress and water balance changes in lowland forest sites. Only if increases in forest area are realised, will national forest productivity be maintained at current levels.

- In terms of optimal use of land capability, secondary growth (extensive forestry) practices would become, by default, the most appropriate land use in the lowland areas due to reductions in land capability for agricultural and pasture uses for these sites.

- Less productive forests could be potentially expanded to cover a wider area under the scenarios considered and potential increases in total production could be possible if secondary forest expansion is allowed to occur. This optimal use of land capability types would only occur if lands currently in agricultural use are allowed to revert back to forest regeneration. Figure 8 depicts the percent of optimal land capability classes under current climate and future scenarios.

- Increasing population pressures, critical food production requirements, expanding banana plantations and continued forest clearing for cattle ranching are obvious obstacles to the return of lowland non-forest areas to extensive forestry practices at present.

Adaptive Management Responses for Costa Rica

Regional scale analysis presented in this pilot project offers an initial assessment of potential change in ecoclimatic zones at a spatial scale which allows for the consideration of possible management responses to climatic change. Two general classes of anticipatory management responses can be developed. These are responses to direct climatic impacts on existing forest areas and responses to potential secondary changes in land use.

In general, management actions in response to an uncertain range of direct impacts on existing forest areas should include the maintenance of forest management options and natural ecosystem resilience to change (Halpin *et al.*, 1991; Secrett, 1992). Direct impacts of climate change such as direct physiological stress, changes in species composition, changes in vegetation structure, changes in forest hydrology and changes in disturbance regimes will be best met by healthy forest systems (Secrett, 1992). In general, the maintenance of natural ecosystem resilience to climatic change could involve the application of management initiatives which promote natural vegetation regeneration and migration. Examples of this type of strategy would be programs which encourage the maintenance of viable populations of species, reduced protected area fragmentation, low impact timber harvesting techniques and the maintenance of migration corridors spanning different habitat and altitudinal gradients. Sound, sustainable forestry practices and forest protection efforts, which would be beneficial under any climatic condition, would be especially useful in anticipation of changing environmental conditions. The possibility of future climatic change strengthens the need to invoke sustainable forest management practices which have already been outlined in the Tropical Forest Action Plan, the Conservation Strategy for Costa Rica's Sustainable Development and numerous existing forest policy mandates (Tosi *et al.*, 1992).

Anticipatory responses to secondary changes in land use and land capability will involve the maintenance of land-use options for the future. A wider range of future management options can be maintained through conservative approaches to natural area land-use which promote flexibility in the type of use over time. For example, the maintenance of multiple-use buffer zones around reserve areas increases the range of possible future land-use interventions if climatic change warrants alterations in the shape, size or location of reserve areas in the future.

A general increase in the time horizon used in assessing land use programmes and forest use zoning would aid in the maintenance of land use options. An increased use of geographic analysis optimising land capability with current spatial relationships to agricultural activities, transportation and population centers could increase efficient natural resource use and reduce risks to climatic impacts.

References

Calvo, J.C. 1990. The Costa Rican National Conservation Strategy for Sustainable Development: Exploring the Possibilities. *Environm. Conserv.* **17(4)**:355-358

Campos Ortiz, M. 1990. *Climate Station Data for Costa Rica.* IMN pers. comm.

Cruz, W. & R. Reppetto. 1992. *Resource Accounting in Costa Rica.* World Resources Institute, Washington D.C.

Davis, M.B. 1989. Lags in vegetation response to greenhouse warming. *Climatic Change* **15**:75-82.

Emanuel, W.R., Shugart, H.H. & Stevenson, M.P. 1985. Climatic change and the broad-scale distribution of terrestrial ecosystem complexes. *Clim. Change* **7**:29-43.

Flores Rodas, J.G. 1985. *Diagnostico del sector industrial forestal.* EUNED-UNED/PEA, DGF/MAG, FPN, Tinker. San Jose, Costa Rica.

Fundacion Neotropica. 1988. *Desarrollo Socioeconomico y el Ambiente Natural de Costa Rica.* San Jose, Costa Rica.

FN/CI (Fundacion Neotropica & Conservation International). 1988. *Costa Rica: Assessment of the Conservation of Biological Resources.* Conservation International, Washington, D.C.

Gomez, L.D. & W. Herrera. 1986. *Vegatacion y Clima de Costa Rica, Vol I EUNED.* San Jose, Costa Rica.

Graham, R.W. 1988. The Role of Climate Change in the Design of Biological Reserves: The Paleoecological Perspective for Conservation Biology. *Conserv. Biol.* **2(4)**:391-394.

Granger, O.E. 1989. Implications for Caribbean Societies of Climate Change, Sea-Level Rise and Shifts in Storm Patterns. In: J. Topping Jr. (ed) *Proceedings of the Second North American Conference on Preparing for Climatic Change: A Cooperative Approach.* The Climate Institute, Washington D.C.

Halpin, P.N., P.M. Kelly, C.M. Secrett & T.M. Smith. 1991. *Climate Change and Central American Forest Systems: Costa Rica Pilot Project (project background report).*

Halpin, P.N. 1992. *Potential impacts of climate change on protected area: global assessments and regional analysis illustrated with a case study in Costa Rica.* Proceedings of the IV World Parks Congress, Caracas Venezuela .

Hansen, J., I. Fung, A. Lacis, S. Lebedef, D. Rind, R. Ruedy, G. Russel & P. Stone.1988. Global Climate Changes as Forcast By the Goddard Institute for Space Studies Three Dimensional Model. *J. Geophys. Res.* **93**:9341-64.

Hartshorn, G. S *et al.* 1982. *Costa Rica Country Environmental Profile: a Field Study.* Tropical Science Center/USAID, San Jose, Costa Rica.

Hartshorn, G.S. 1983. Plants. In: Janzen, D. (ed) *Costa Rica Natural History.* Univ. of Chicago Press, Chicago.

Holdridge, L.R. 1949. Determination of World Plant Formations from Simple Climate Data. *Science* **105 (2727)**:367-368.

Holdridge, L.R. 1959. Simple method for determining potential evapotranspiration from temperature data. *Science* **130**:572.

Holdridge, L.R. 1967. *Life Zone Ecology.* Tropical Science Center, San Jose, Costa Rica.

Holdridge, L.R., W.C. Grenke, W.H. Hatheway, T. Liang & J.A. Tosi Jr. 1971. *Forest Environments in Tropical Zones: A Pilot Study.* Pergamon Press, Oxford, 749pp.

Holdridge, L.R. & J.A. Tosi Jr. 1972. *The World Life Zone Classification System and Forestry Research.* Seventh World Forestry Congress. 7CFM/C:V2G FAO Rome: 4826-4873.

Hunter, M.L. G.L. Jacobson Jr. & T.Webb, III. 1988. Paleoecology and the Coarse Filter Approach to Maintaining Biological Diversity. *Conserv. Biol.* **2(4)**:375-385.

IN/CR (Instituto Geographia de Costa Rica). 1987. *Mapa Fisico-Politico.* San Jose, Costa Rica.

IDB. 1987. *Forest Development Pilot Project (CR-0070) Project Report PR-1583-A.* Washington, D.C.

IMN - Instituto Meterologico Nacional. 1985. *Climas de Costa Rica.* San Jose, Costa Rica.

Janzen, D. 1983. *Costa Rica Natural History.* Univ. of Chicago Press, Chicago.

Kelly, P.M. 1991. *Regional Climate Change Scenarios for Costa Rica.* Climatic Research Unit, Univ. of East Anglia. (unpublished report).

Leemans, R. & W.P. Cramer. 1990. The IIASA database for mean monthly values of temperature, precipitation and cloudiness on a global terrestrial grid. WP-90-41, International Institute for Applied Systems Analysis, Laxenburg, Austria, 61pp.

Leemans, R. and Halpin, P.N. 1992. Biodiversity and global change. In: J. McComb (ed) *Biodiversity Status of the Earth's Living Resources.* World Conservation Monitoring Center, Cambridge, U.K.

Mackintosh, G; J.Fitzgerald & D. Kloepfer. 1989. Preserving Communities and Corridors. Defenders of Wildlife, G.W. Press

Manabe, S. & R.T. Wetherald. 1987. Large scale changes in soil wetness induced by an increase in carbon dioxide. *J. Atm. Sci.* **44**:1211-1235.

Maul, G.A. 1989. Implications of Climatic Change for the Wider Caribbean Region: An Overview. In: J.Topping Jr. (ed) *Proceedings of the Second North American Conference on Pre-*

paring for Climatic Change: A Cooperative Approach. The Climate Institute, Washington D.C.

Mitchell, J.F.B. 1983. The seasonal response of a general circulation model to changes in $C0_2$ and sea temperatures. *Q. J. Roy. Met. Soc.* **109**:113-152.

O'Brien, S., B.P. Hayden & H.H. Shugart. 1992. Global climatic change, hurricanes and a tropical forest. *Clim. Change* **22**:175-190

PAFT-CR. 1990. *Plan de Accion Forestal para Costa Rica. Symposium Document.* San Jose, Costa Rica

Peters, R.L. & J.D. Darling. 1985. The Greenhouse Effect and Nature Reserves. *Biosci.* **35** **(11)**:707-717.

Portig, W. 1975. The Climate of Central America. In: Shwerdtfeger (ed) *Climates of Central and South America; World Survey of Climatology Vol. 12.* Elsevier.

Prentice I.C., R.S. Webb, M.T. Ter-Mikhaelian, A.M. Soloman, T.M. Smith, S.E. Pitovranov, N.T. Nikolov, A.A. Minin, R. Leemans, S. Lavorel, M.D. Korzukhin, H.O. Helmisaari, J.P. Hrabovszky, S.P. Harrisson, R.W. Emanuel, & G.B. Bonan. 1989. *Developing a global vegetation dynamics model: results of an IIASA summer workshop.* IIASA Research Report RR-89-7. August 1989. IIASA, Laxenburg, Austria.

Prentice, I.C., W. Cramer, S.P. Harrison, R. Leemans, R.A. Monserud & A.M. Solomon. 1992. A Global Biome Model Based on Plant Physiology and Dominance, Soil Properties and Climate. *J. Biogeogr.* **19**:117-134.

Sader, S.A. & A.T. Joyce. 1988. *Deforestation Rates and Trends in Costa Rica.*

Sawyer, S.A. & A. Lindsey. 1971. Vegetation Life Zones in Costa Rica. *Indiana Acad. of Science Monogr.* **no. 2.**

Secrett, C.M. 1991. *Costa Rica Forest Resources Overview.* International Institute for Environment and Development Project Paper, London.

Secrett, C.M. 1992. *Adapting to Climate Change; A Strategy for the Tropical Forest Sector.* International Institute for Environment and Development Project Paper, London.

Schlesinger, M. & Z. Zhao. 1988. *Seasonal climatic changes induced by doubled CO_2 as simulated by the OSU atmospheric GCM/mixed layer ocean model.* Oregon St. U., Corvallis, OR, Climate Research Institute.

Smith, T. M., H.H. Shugart & P.N. Halpin. 1990. *Global Forests, in Progress Reports on International Studies of Climate Change Impacts.* USEPA Washington D.C.

Tosi, J.A. Jr. 1969. *Mapa Ecologico Republica de Costa Rica.* Tropical Science Center, San Jose, Costa Rica.

Tosi, J.A. *et al.* 1988. *Manual para la determinacion de la capacidad de uso de las tierras de Costa Rica.* Tropical Science Center, San Jose, Costa Rica

Tosi, J.A., Watson, V. & J. Echeverria. 1992. *Potential Impacts of Climate Change on the Productive Capacity of Costa Rican Forests: A Case Study.* Tropical Science Center, San Jose, Costa Rica.

TSC/WRI (Tropical Science Center and World Resources Institute). 1991. *Costa Rica Natural Resources Accounting Study.* San Jose, Costa Rica

Walter, H. 1985. *Vegetation of the Earth and Ecological Systems of the Geo-Biosphere (Third Edition).* Springer-Verlag, Berlin

WB/UNDP (World Bank/United Nations Development Program). 1990. *Forestry Residues Utilization Study: Costa Rica.* WB/UNDP Bilateral Aid Energy Sector Management Assistance Program, Activity Completion Reports No. 108A/90 and 108B/90. Washington D.C.

Modelling vegetation dynamics of miombo woodlands of Zambezian Africa under changing climatic conditions

Paul V. Desanker

School of Forestry & Wood Products, Michigan Technological University
1400 Townsend Drive, Houghton, MI 49931,USA;

I. Colin Prentice

Department of Plant Ecology, University of Lund
O. Vallgatan 14, S-223 61, Lund, Sweden.

Introduction

Tropical forests in Africa are environmentally critical due to their role in protecting water catchment areas and controlling soil erosion, and they provide vital renewable resources for construction, grazing, fuelwood, valuable timber for commercial use, and other traditional uses like herbal medicines. As climate changes, and use of these resources intensifies with growing population, the forest resources will continue to change rapidly. The limited financial capabilities of most African countries, and their economic dependence on natural resource production, make them especially vulnerable to the impacts of global climate change.

Natural resources in many countries of Africa are already under considerable stress due to rapid population growth and such phenomenon as desertification and deforestation. Long term grazing, for instance, leads to desertification of formerly productive land (Schlesinger *et al.*, 1990). Shukla *et al.* (1990) assessed the effects of Amazon deforestation on the regional and global climate by using a coupled numerical model of the global atmosphere and biosphere. They found that there was a significant increase in predicted surface temperature and a decrease in evapotranspiration and precipitation when the tropical forests in the model were replaced by grass. In their simulation, the length of the dry season predicted also increases; such an increase could make reestablishment of the tropical forest after massive deforestation particularly difficult. Factors like these, place forests and people in developing regions of Africa under great risks as stress on natural resources increases. A model developed for the tropical forests in Africa will help assess potential impacts of climate change in these poorly studied but environmentally critical regions, aswell as provide a starting point for further model developments, and data collection. This information will help tropical countries prepare for effects of these changes.

There is extensive experience in modelling dynamics and the response of temperate and boreal forests of North America and Europe to various environmental factors and climate. Forest gap model simulators form an important portion of this work (eg. Doyle, 1981; Solomon *et al.*, 1981; Shugart & Noble, 1981; Shugart *et al.*,1981; Shugart & West, 1977; Solomon *et al.*, 1980; Leemans & Prentice, 1987; Pastor & Post, 1986). Gap or forest succession models simulate changes in composition and structure on patches of 400-1,000 m^2 by simulating mixed species tree populations and their interactions with the environment through changes in availability of light, water, and nutrients. Canopy gaps created by death of a tree are believed to be crucial in forest dynamics (Platt & Strong, 1989). The concept of the forest gap or *gap phase* is attributed to Watt (1947). The gaps become localised sites of regeneration and subsequent growth. Differences in gap sizes result in specific vegetation patterns of species based on their shade tolerance characterisation (Canham, 1989; Whitmore, 1989). In tropical forests, small gaps release suppressed saplings of climax species, while in large gaps pioneer species can also establish (Swaine & Whitmore, 1988). The mature forest ecosystems are seen as the relatively consistent average of the responses of the dynamics of such gaps (Shugart, 1984).

In Africa, a gap model has been used to describe effectively the dynamic properties of forests in the Cape, South Africa, by Van Daalen and Shugart (1989). The model used in the Cape, as well as other models for forests elsewhere (see Shugart, 1984), vary in details of processes and parameter requirements. They can be parameterised from a knowledge of basic physiology, morphology, or forestry. There are two main implementations of gap models, namely the JABOWA/FORET type (Botkin *et al.*,1972; Shugart & West, 1977; Shugart, 1984), and the FORSKA model (Leemans & Prentice, 1987; Prentice & Leemans, 1990). The FORSKA model does not depend strongly on age related data, and in general, has fewer parameters than other gap models, making its formulation particularly attractive in tropical regions where data are limiting.

In this paper, we report on an application of FORSKA Version 1 in developing MIOMBO, a model for the miombo woodlands of the Zambezian region of Central Africa. We parameterise a base case, towhich we apply a drought routine to simulate the differential development of various species of *Brachystegia*. We use site specific information on climate and climate change scenarios for the central region area of Malawi, around Lilongwe (13°58'S, 33°42'E).

Background

The vegetation of Africa

The vegetation of Africa is classified by White (1983) using a system that is primarily physiognomic, with units grouped according to the floristic regions (phytochoria) in which they occur. Physiognomy covers all aspects of the structure of vegetation, but most classifications rely heavily on a few characters such as height, density, thorniness, deciduousness, and so on. White (1983) recognised 16 major vegetation types, or formations, at the highest rank characterised by their physiognomy. The first two formations of regional extent are forest, and woodland. A forest is defined as a continuous stand of trees at least 10 m tall, their crowns interlocking, and a woodland as an open stand of trees at least 8 m tall, with a canopy cover of 40% or more, and with the field layer usually dominated by grasses.

Regional variants of the main formations are designated by the phytogeographical area in which they occur. Eighteen major phytochoria are recognised for Africa and two for Madagascar. The phytochoria are based on their richness (or otherwise) of their endemic floras at the species level. At the rank of Region, four fundamentally different types of phytochoria are recognised (White, 1979). These are: i) Regional Centres of Endemism; ii) Archipelago-like Centres of Endemism; iii) Archipelago-like Centres of Extreme Floristic Impoverishment; and iv) Regional Transition Zones and Mosaics. The Regional Centres of Endemism are separated by transition zones. The transition zones show a simple gradual replacement of one flora by another which is little complicated by endemism. A brief description of the woodlands of the Zambezian Regional Center of Endemism is summarised below from White (1983), from which further details about all the Regions and vegetation types of Africa can be found.

The Zambezian Regional Centre of Endemism

The Zambezian Region as described by White (1983) extends from 3°S to 26°S and from the Atlantic Ocean almost to the Indian Ocean. It includes the whole of Zambia, Malawi Zimbabwe, large parts of Angola, Tanzania and Mozambique, and smaller parts of Zaire (Shaba), Namibia, Botswana, and South Africa (Transvaal) (Area: 3,770,000 km^2.) Most of this is woodland. According to Ernst (1971), this corresponds to the limits of resistance to heat and cold of the main trees (+48°C for the leaves and buds, -2°C for the stems of *Brachystegia spiciformis*).

The larger part of the Zambezian Region is occupied by the Great African Plateau and lies more than 900 m above sea level, rising in places near the rim to over 2,500 m. The higher ground supports Afromontane communities. Most of the Zambezian Region is drained by the Zambezi River but the northern fringes lie within the catchment area of the Zaire.

Almost the whole of the Region lies within Walter's tropical summer-rainfall zone. Except towards the coast the climate is continental in character with an appreciably larger seasonal variation in temperature than that of the Guineo-Congolian Region. There is a single rainy season, chiefly from November to April, sometimes interrupted by a dry spell lasting two to three weeks. Rainfall is between 500 and 1,400 mm per year and in general decreases from north to south, but there are pronounced variations. Mean annual temperature, which varies from 18°C to 24°C, is correlated more with altitude than latitude. The Zambezian Region, after the Sahara, is the largest major phytochorion in Africa. The literature dealing with its vegetation has recently been reviewed by Werger and Coetzee (1978). White (1983) describes the various vegetation types in this Region. Most of the Zambezian is woodland, representing the climax in many places, while it is secondary in others, or has been profoundly modified by cultivation and fire. Three main types of woodland, namely miombo, mopane and undifferentiated Zambezian woodland are shown on the UNESCO/AETFAT Vegetation Map of Africa. In this paper, we shall restrict our attention to the miombo woodlands.

Miombo Vegetation

Description of miombo vegetation

Miombo is a type of woodland dominated by *Brachystegia*, either alone or with *Julbernadia* or *Isoberlinia*. The most charactersitic miombo soils are leached and acid. They are often shallow and stony, or the rooting environment is otherwise restricted by the occurrence of laterite or a gley horizon near the surface. On the drier and shallowerplateau soils the original vegetation was almost certainly miombo but on the moister and deeper soils in the higher-rainfall areas, miombo has probably replaced dry evergreen forest or transition woodland following cultivation and fire. The trees have interlocking or nearly interlocking crowns, but the foliage is light, the branches being most often extended in umbrella fashion, so that the forest is light and airly. Nineteen species of *Brachystegia* and three species in related genera, namely *Julbernadia globiflora, J. paniculata* and *Isoberlinia angolensis*, occur in miombo as dominants. Miombo is mostly 10-20 m tall but scrub miombo can be as short as 3 m. On certain deep soils, *Brachystegia spiciformis, B. longiflora*, and *B. utilis* reach a height of 30 m and the climax in such situations is possibly dry evergreen forest or transition woodland rather than miombo.

Most miombo woodlands are semi-deciduous, but some are completely deciduous and some are almost evergreen. *B. taxifolia* is evergreen while all other Zambezian *Brachystegia* species in high-rainfall areas are semi-deciduous (ie. briefly deciduous, old leaves are shed as new leaves unfold some weeks or even months before the end of the dry season). Of the associated species, a few, like *Diospyros batonica* are completely evergreen, and some like *Pterocarpus angolensis* are completely deciduous. In areas of high temperatures and lower rainfall, miombo is completely deciduous for up to two months.

Moisture requirements

White (1983) distinguishes between wetter miombo and drier miombo using the 1,000 mm mean annual rainfall isohyet to seperate the two, while Chidumayo (1987) uses 1,100 mm in Zambia because it approximates the 1.1 aridity ratio (mean annual precipitation/annual potential evapotranspiration) line.

Wetter miombo: Rainfall usually more than 1,000 mm per year, but less on Kalahari sand. Canopy height often 15 m. Floristically rich. Includes nearly all the miombo dominants. *Brachystegia floribunda, B. glabernima, B. taxifolia, B. wangermeeana* and *Marquesia macroura* are widespread. Associated vegetation includes dry evergreen forest and thicket, swamp forest, evergreen riparian forest, and wet dambos.

Drier miombo: Rainfall less than 1,000 mm. Canopy height usually less than 15 m. Floristically poor. *Barchystegia floribunda* etc. absent or very local. *Brachytegia spiciformis, B. boehmii* and *Julbernadia globiflora* are often the only dominants present. Associated vegetation includes dry deciduous forest and thicket, deciduous riparian forest, and dry dambos.

Soil preferences

Malaisse (1978) lists the following soil preferences and rooting characteristics of the main species in Zambia from research there:

Julbernardia paniculata is very adaptable and occurs on all sites.
Brachystegia spiciformis prefers really deep soils and cannot stand waterlogging.

B. longifolia prefers deep sandy soils, cannot penetrate murram.
B. floribunda prefers heavy textured soils and can penetrate murram.
B. boehmii prefers clay loams and can penetrate murram.
Isoberlinia angolensis is very adaptable and lives within a permanent (moving) watertable.

Fire and regeneration dynamics

Lawton (1978) develops the relationships between miombo and mateshi, and between miombo and chipya, based mainly on tolerance of fire and shade. The term *chipya* (meaning "burnt") was applied by Trapnell (1953) to vegetation in which various trees other than *Brachystegia, Julbernardia,* and *Isoberlinia* grow mixed in very tall grass. Such vegetation burns fiercely and the trees are remarkedly fire resistant. Chipya occurs on sites formerly occupied by forest or transition woodland and owes its existance to cultivation and fire. Mateshi, on the other hand, is a dry evergreen forest with a closed canopy at 15-25 m. Chipya is believed to have been derived from mateshi by opening up the canopy for cultivation, and by burning in the dry season (Lawton, 1978), and chipya can revert to mateshi if burning is pevented (Trapnell, 1953; Lawton, 1964, 1978).

Based on burning experiments reported by Trapnell (1959), Lawton (1978) defines five main ecological groups, which we refer to as *species-fire roles*:

(1) fire-hardy species intolerant of shade;
(2) species with moderate fire-resistance; able to invade a light canopy formed by woody plants 3-4 m tall; but unable to invade or persist among the tall grasses of chipya;
(3) species able to invade under those of group (2), but not those of group (1), and needing some protection from fire if they are to grow through the sapling stage to form mature trees;
(4) fire-sensitive species, usually evergreen, unable to withstand dry season fires;
(5) ubiquitous species tolerant of fire and shade.

Chipya species form group (1); and when they form a light canopy, they create conditions favourable for group (2), the Uapaca group. Under the protection of group (2), the miombo canopy species of *Brachystegia* and *Julbernardia* (group 3) are able to pass from sapling stage to form a mature woodland canopy. Group (2) also protects the regeneration phase of the fire-sensitive evergreen trees of the dry evergreen forest that form group (4). Group (5) consists of a number of ubiquitous species with a wide ecological range.

Fire occurrence

Lawton (1978) discusses fire in miombo: fire is almost always started by man, with the intensity of dry-season fires largely a function of the degree of development of the herb layer, and this in turn is dependent upon the nature of the tree canopy. When the first three groups (above) develop canopies that suppress vegetation, intensity of dry-season fires is reduced, and the next group can invade. Fires in the late dry season are more destructive, as this time new leaves have emerged, and new coppice growth is burnt back to the ground level and crowns of canopy trees may be damaged. Burning experiments carried out in Zambia (Trapnell, 1959; Lawton, 1972) and Zaire (Delvaux, 1958; Symoens & Bingen-Gathy, 1959) identified fire-tolerance of constituent species of miombo. Among the canopy dominant species, *Brachystegia* and *Julbernardia* are considered fire-tender species, while *Isoberlinia angolensis* is semi-tolerant. Hopkins (1965) reported an average loss of 7.9 % in stocking per year on an area subjected to late burning, with survival increasing with height and basal area. Glover (1968) found trees 1-4 m high to be the most vulnerable to fire.

The model

The FORSKA model is based on the basic principles of the individual tree-based gap models. Such models began with the JABOWA model of Botkin *et al.*, (1972), from which a family of forest succession models is descended (Shugart, 1984). FORSKA Version 1 is fully documented in Leemans and Prentice (1989) and the model is described in Prentice and Leemans (1990). Here, we only outline the principles and emphasise the changes made over the documented version.

Gap models simulate a forest stand by computing establishment, growth and mortality processes in independent patches. Establishment and mortality are treated as stochastic processes, while growth is deterministic. In MIOMBO, each simulated tree is described by its species, diameter at breast height (dbh), leaf area, and bole height. The environment of each patch is calculated from the tree characteristics via the vertical profile of leaf area index (LAI) and the total patch biomass. Site variables are held constant.

Regeneration

A random number of saplings of all species is allowed to regenerate, with light conditions as the only restriction for a species to establish. If light intensity is above the compensation point for the species, new saplings are regenerated for that species from a Poisson distribution with expected number of seedlings reduced by a shading factor, SHF, where SHF is defined as follows:

$$SHF = ((kI_f - c)/(kI_f + a - c)) / ((kI_o - c) / (kI_o + a - c)) \text{ check}$$

where

I_f = forest-floor light intensity
k = extinction coefficient
c = compensation point
a = half-saturation point
I_o = above-canopy light intensity.

New saplings appear with a height of 1.3 m (breast height), a dbh of 1 cm plus a random fraction of the dbh increase which the sapling would achieve in a full time step, and an assigned initial leaf area. Bole height is initially zero.

Growth

Light intensities are averaged over the growing-season and patch, and light extinction within the canopy follows the exponential decay curve with LAI as the exponent (the Beer-Lambert Law). A species compensation point determines the species ability to grow at very low PAR, while its half-saturation point is its growth efficiency in the lower PAR range.

Tree height is a function of dbh following the Mitscherlich curve (Meyer, 1940), with height approaching maximum height as dbh becomes large.

In earlier gap models (discussed in Shugart, 1984), foliage was concentrated at the top of the crown. In FORSKA, foliage is distributed vertically along the crown depth as most boreal species (for which FORSKA was initially constructed) are deep crowned. Leaf area of each tree is distributed uniformly between top and bole heights. A parameter was introduced in MIOMBO, called bole height factor (BHF), that relates the minimum allowable ratio of bole height to the total tree height. The BHF controls the depth of the crown relative to the total tree height, and so, simulates flat to cylindrical crown shapes, by concentrating leaf area along shallow or deep crowns. For example, flat crowned trees would have a small crown depth, and so a relatively longer bole height than oval crowned trees. Since gap models are spatially explicit in the vertical direction, longer bole heights would concentrate the leaf area of the tree in the upper fraction of total tree height.

Each layer of 0.5 m of the crown contributes to net assimilation, and incurs a maintenance cost proportional to the sapwood volume needed to supply it. Net growth is reduced in proportion to patch biomass. Biomass itself is calculated as the sum of dry weights using a factor that converts wood volume to biomass weight.

Destructive Effects of Fire

Fires are believed to be more destructive when lots of fuel (litter) accumulates and grass is plentiful. Grasses are more abundant when the canopies are light, that is, when LAI is low. So when more light is reaching the forest floor, the more the grasses, and the greater chance of a fire being destructive. In MIOMBO, we generate (destructive) fire and its effects as follows: Calculate amount of shading at the forest floor as I(f)/I(0), to reflect forest conditions favoring grasses. Then taking an average frequency of destructive fires per year of 8% (Hopkins, 1965), reduce this frequency by the amount of shading to get a probability of fire as

$$P(Fire) = Fo * 0.08 * I(f)/I(0)$$

where Fo is a parameter to be varied during simulation to suppress or increase fire occurrence.

Resistance to the fire is represented by each species' fire tolerance. Small trees (dbh < 5 cm) of fire-intolerant species are killed during a destructive fire. For semi-tolerant species, rate of mortality is increased by 0.2. Fire-hardy species are not affected. The rationale for this being that fires in miombo ignite almost every year, such that fuels do not accumulate to large enough amounts to cause total damage in a stand. A more complex fire routine can be used when data are available.

Mortality

Mortality occurs at a low, *intrinsic* rate in vigorous trees and at a much higher rate in suppressed trees. In the present version of MIOMBO, dead trees are not allowed to coppice. This was left out as there was not adequate information on coppicing for all the species in the simulations. Coppicing is very important in the regeneration of miombo, and data will be sought to include it in future versions of the model. Death can also be due to destructive fires as discussed in the preceeding section.

Effects of Moisture Stress

Soil moisture is the limiting factor for tree growth over most of Malawi (Hardcastle, 1977). Hardcastle (1977) computed average stree periods (weeks per year) for temperature-rainfall (*silvicultural*) zones over the whole of Malawi. These range from 15 weeks to 35 weeks at a 7-inch available water capacity. Drier miombo are most common in zones with rainfall less than 1,100 mm, while the wetter miombo are common in zones with higher rainfall. This information was used to derive a moisture stress multiplier, MOIST, as

$$MOIST = 1.0 - a\ (X - 15)$$

where X is the number of weeks of moisture stress in the range 15 to 40, and a is the slope parameter, taken to be 0.10 for wetter miombo, and 0.05 for drier miombo.

Results and discussion

Model Parameters

DeAngelis *et al.*, (1981), Malaisse (1978), White (1983), Ernst (1988), Stromgaard (1985), and Chidumayo (1987, 1988a, 1988b, 1990, 1991) give various characteristics of miombo vegetation or constituent species of miombo. These were used to estimate parameters of the MIOMBO model. The estimated parameter values are given in table 1. Where direct measurements were not available, approximations within suitable ranges were used. These are indicated in table 1. These approximations are reasonable since exact values of most of the parameters are not so important as shown by a sensitivity analysis of FORSKA Version 1 by Leemans (1991).

Base Model Runs

The model was run for 200 years from a bare patch. Figure 1 shows basal area with age for a typical run under non stress conditions. Understorey species (eg. *Pericopsis angolensis*, *Syzygium species*, and *Uapaca* species) dominate at first as forest cover increases, then various overstorey species of *Brachystegia* as well as *Marquesia* and *Julbernardia* slowly overtop the understorey species. In the present formulation, all species have equal chance of establishment, providing light conditions are suitable. Under non stress conditions, the wetter miombo species (*Marquesia* and *B. spiciformis* in these formulations) continue to increase in size and contribute more to basal area after the drier species have reached their upper limit. A more detailed regeneration routine would include moisture requirements by different species, and will be attempted in future when more information is available.

Figure 1.
Basal area with age for a typical run under non-stress conditions (15 weeks of moisture stress per year).

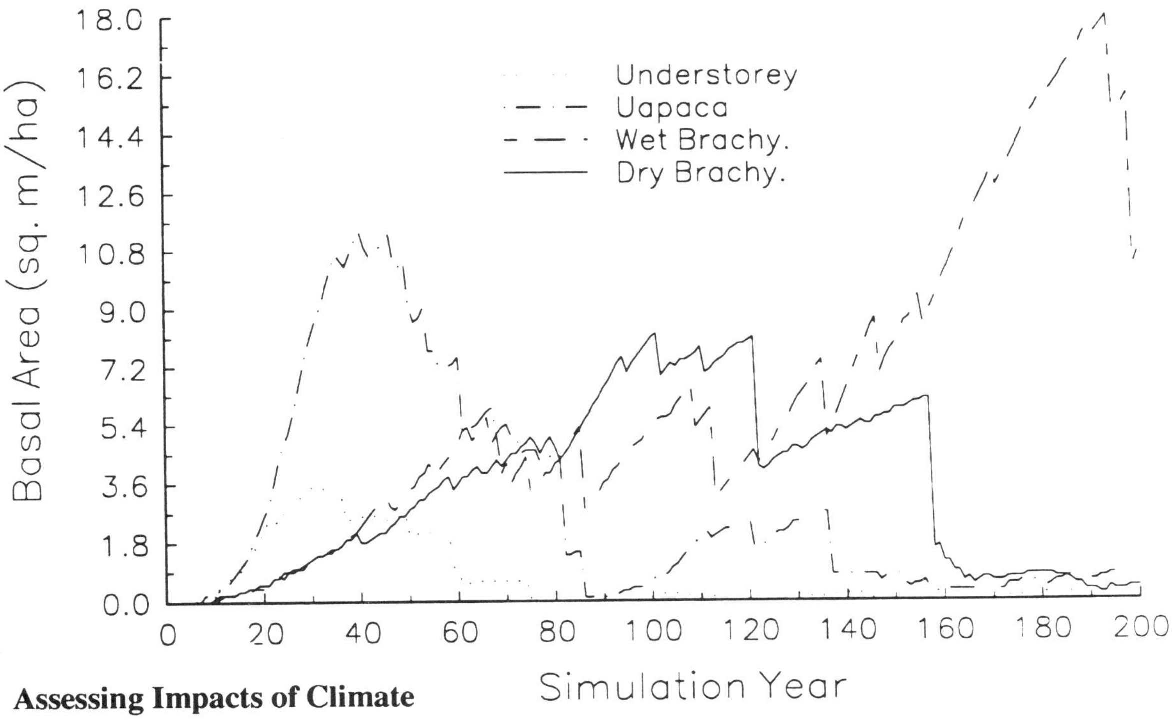

Assessing Impacts of Climate

The model was run under different climate scenarios for Lilongwe, Malawi (Zone D in Hardcastle's classification; average climate data given in Muller, 1982). A climate change scenario over Lilongwe was derived from results obtained with the GDFL general circulation model for a doubling of atmospheric CO_2 (Wetherald & Manabe, 1986), where the normal climate was taken from Leemans and Cramer (1991). The DFL scenario projected a decrease in annual precipitation of about 40%, however, with an increase in mean annaul temperature of 3.2°C. The seasonal pattern of rainfall and temperature changed: present rainfall begins in November and ends in March, with most rain in Dec-Feb; under the GDFL scenario, rainfall started in November as before, but was followed by a totally dry month of December, then rain returning in January until March as at present (see Figure 2). (An OSU scenario is also shown, but was not used in this analysis as it produced rather large increases in precipitation. December alone received about 1,000 mm!). The GDFL scenario also resulted in the hottest month of the year shifting from October to November (see Figure 3). The overall effect of the projected GDFL climate change would be to lengthen the period of moisture stress especially in the first half of the growing season. The severe drought that would result shortly after the rainy season begins, would most likely have adverse effects on tree establishment, as well as on crops that are typically planted at the onset of rains in November/December. These changes in the growing season conditions were incorporated into the MIOMBO model by increasing the period of moisture stress from 15 to 21, and to 24 weeks. The model was then run for 200 years from a bare patch. The wetter miombo was never able to establish under the stressed conditions due to the GDFL scenario, with basal area staying below 1 m^2 for the length of the simulation, compared to a steady increase to about 8 m^2 per hectare under normal conditions. After about a hundred years, the drier miombo was clearly dominating, having outcompeted the wetter

Table 1.
Summary of the Model

[a] State Variables

Description	Symbol
Species	TAX
Diameter	DBH
Bole Height	HBC
Leaf Area	LEA
Age	AGE
Dead Tree Species	DTAX
Dead Tree Diameter	DDBH
Dead Tree Age	DAGE
Number of Dead Trees	NUM

[b]Input Functions

Description	Equation	Symbol
Sapling establishment probability	P(n)	
Initial leaf area for saplings	Li	
Tree height	H	
Bole height	B	
Biomass	W	
Available light in canopy	Iz	
Net assimilation reduction factor	Pz	
Growth	$d'(D^2H)$	
Leaf area	L	
Mortality rate	X	
Probability of death	PD	
Probability of destructive fire	PFIRE	
Sapwood maintenance cost		DEL
Light compensation point	C	LCP

[c]Parameter values (independent of species)

Symbol	Name	Definition	Unit	Range
INS	Io	Growing season-average light intensity	umol/m²/s	980
KCO	k	Light extinction coeffient		0.4
WMX	Wmax	Site maximum stem biomass	Mg/ha	300-600
PAT		Patch size	ha	0.1-1.0
NPAT		Number of patches		5-10
TIM	T	Time step	yr	2
NTIM		Number of timesteps		200
VINC	dz	Vertical integration step through canopy	m	0.1
DHB0	Do	Minimum diameter	cm	1.0
RHO	p	Steepness parameter for mortality		999
DELC		Sapwood maintenance cost coefficient		0.02-0.09
LCF		Light compensation -half saturation coefficient		0.1
TDI		Threshold relative growth efficiency		0.025
UMX		Suppressed mortality rate	/yr	0.46
LAF		Sapwood turnover rate	/yr	0.004
NTAX		Number of species		14

[d] Species Specific Parameters

Symbol	Name	Definition	Unit	Range
NAM		Name of species		
HMX	Hmax	Maximum species height	m	8-30
ALP	a	Half-saturation point	umol/m²/s	100-300
HDS	s	Initial slope of dbh-Ht	m/cm	0-3
EST	E'	(total)Sapling establishment	rate/ha/yr	20-30
UMN	Uo	Intrinsic mortality rate	/ha/yr	
LAC	C	Initial leaf area to D2 ratio		0.05-0.4
BHF	BHF	Minimum bole to total height ratio		0.4-0.7
FTOL		Tolerance to fire		{1,2,3}
FU0		Fire mortality rate		0.20

Simulation with MIOMBO.F for Miombo Woodland in Lilongwe, Malawi

Miombo Woodlands with Moisture Effects 24th August 1991

0.4	980.0	400.0	0.03	0.00	0.46	21.0
0.1	5	2.0	50			
0.5	1.0	999.				
0.02						
14						

Diplorhynchus condylocarpon									
10.0	0.9	600	4	0.033	0.100	0.50	1	0.0	2
Hymenocardia acida									
12.0	1.2	600	4	0.033	0.100	0.50	1	0.0	2
Pterocarpus angolensis									
14.0	1.0	600	4	0.046	0.100	0.50	1	0.0	2
Syzygium guineensis									
7.0	0.7	600	4	0.046	0.100	0.40	1	0.0	2
Vitex doniana									
7.0	0.8	600	4	0.056	0.100	0.40	1	0.0	2
Uapaca									
14.0	1.2	450	4	0.056	0.100	0.40	2	0.10	2
Brachystegia floribunda									
21.0	0.8	450	2	0.025	0.080	0.70	3	0.20	1
Brachystegia glabernima									
21.0	0.8	450	2	0.025	0.080	0.70	3	0.20	1
Brachystegia taxifolia									
21.0	0.8	450	2	0.025	0.080	0.70	3	0.20	1
Brachystegia wangermeeana									
21.0	0.8	450	2	0.025	0.080	0.70	3	0.20	1
Marquesia macroura									
21.0	0.8	450	2	0.025	0.080	0.70	3	0.20	1
Brachystegia spiciformis									
21.0	0.8	450	2	0.025	0.080	0.70	3	0.20	2
Brachystegia boehmii									
21.0	0.8	450	2	0.025	0.080	0.70	3	0.20	2
Julbernadia globiflora									
21.0	0.8	450	2	0.025	0.080	0.70	3	0.20	2

0	NO STAND INITIALISATION
0	NO FILE WITH OUTPUT ARRAY
5	OUTPUT INTERVAL (TIMES TIMESTEP)
0	NO SEPERATE DECRIPTION OF PATCHES
	5.0 START OUTPUT AFTER 10.0 YEARS

Figure 2.
Average monthly precipitation for Lilongwe, Malawi, from different datasets: Muller, IIASA database by Leemans and Cramer (1991), Hardcastle (1977), and from the GFDL and OSU general circulation models.

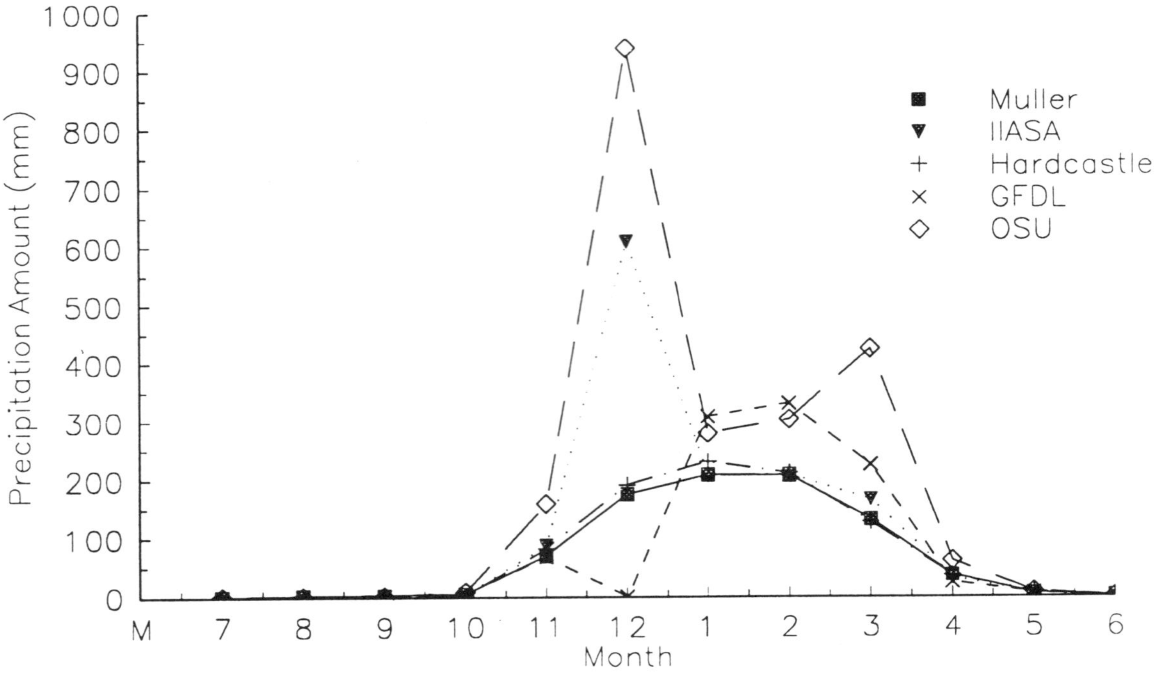

Figure 3.
Average monthly temperature for Lilongwe, Malawi, from different datasets as in figure 2.

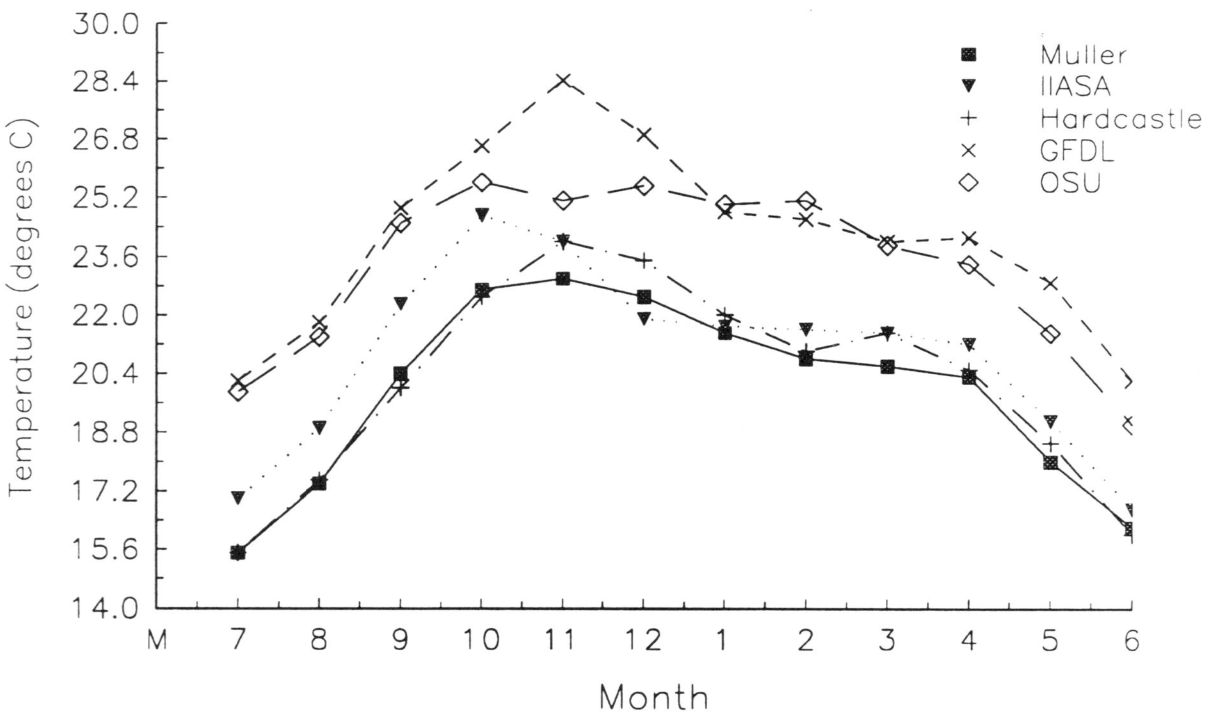

miombo. Under no stress conditions, the drier miombo decreased in influence as the wetter miombo was able to thrive. Tree densities for the three levels of stress are shown in figure 4, while figure 5 shows the pattern in density for the drier miombo.

The seasonal patterns in rainfall and temperature are crucial in defining beginning and end of the growing season, as well as defining the quality of that growing season. Changes in climate will be important both in terms of total annual changes, as well as seasonally. A disruption of the establishment phase of different tree species will likely have a more dramatic effect on the vegetation than the overall annual changes discussed in this modelling exercise. By using number of weeks of stress for a whole year, we have avoided the assumption that the general circulation model produced reasonable monthly values of rainfall.

Gap models require relatively few parameters, and will be useful in many forests of Africa where data on all species of a forest are limited. The physiological basis of FORSKA Version 1 (and so of MIOMBO), enabled parameters to be estimated from limited data by using typical values for types of species (in terms of shade tolerance, crown shape, size, etc.). The MIOMBO model has yet to be properly validated with measurements from specific forests. Therefore, the results should be interpreted with care. As it is, deviations from the base can indicate how miombo forests are likely to respond to imposed changes.

Figure 4.
Tree density for 4 species groups for 15, 21 and 24 weeks of moisture stress (all simulations started from a bare patch).

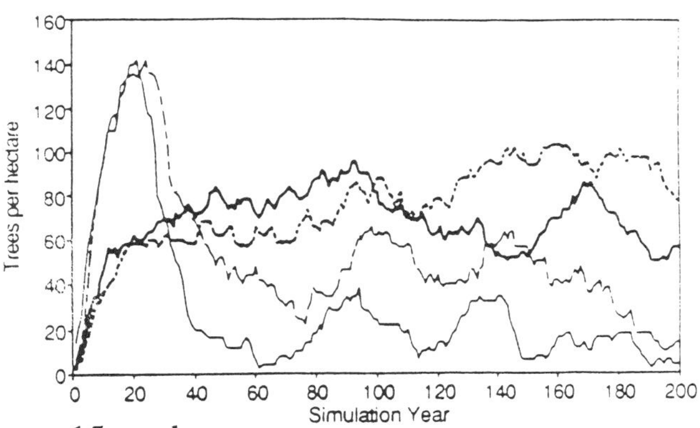

15 weeks

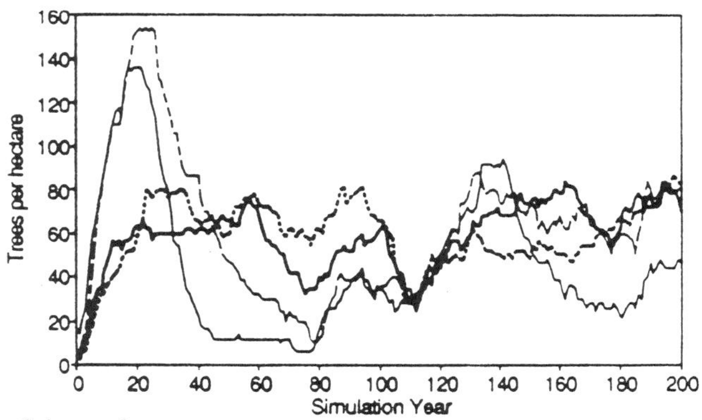

24 weeks

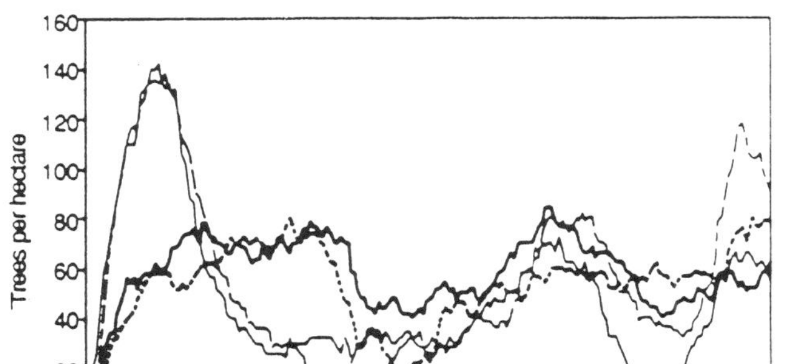

21 weeks

Figure 5.
Tree density for 15, 21 and 24 weeks of moisture stress for dry *Brachystegia* species.

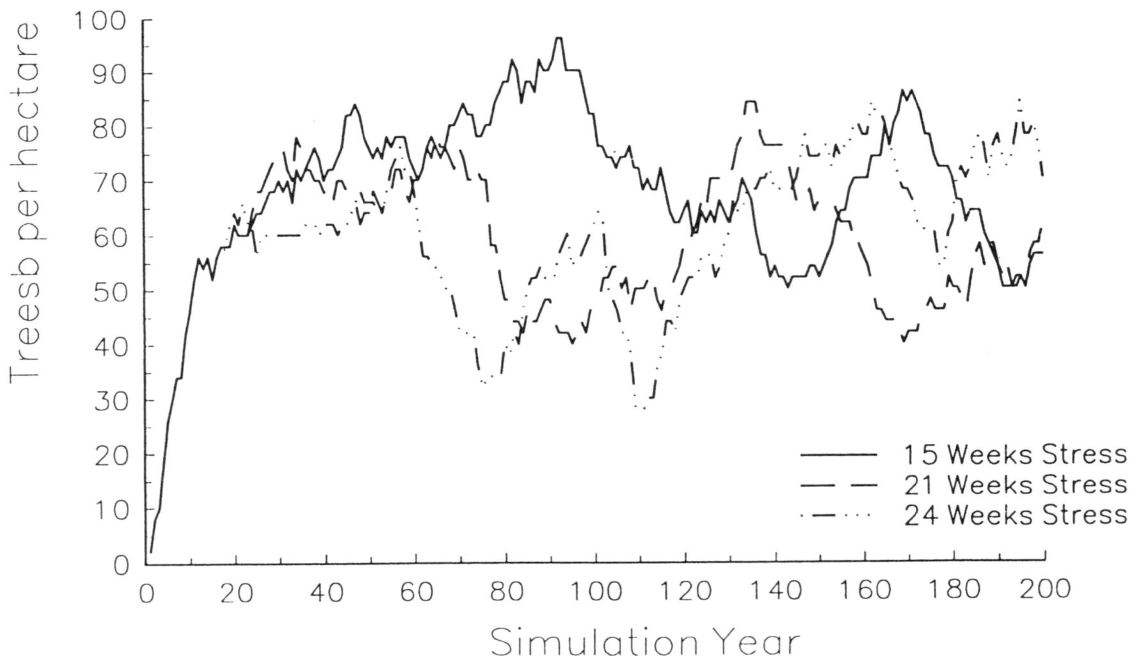

Extensive data covering all component species of miombo (and other forest types of Africa) are lacking. The little data available are mostly on commercial spècies. Collection and publication of original results on measurements of all species within many vegetation types will be useful in studying forest biomass relations and other dynamics of the forest system across the different Regions of Africa, for example, species limits and migration. Forest biomass is important in soil conservation, and commercial as well as non-commercial species, contribute to this. The non-commercial species often form the major landcover in miombo, and provide a much needed soil cover from the hot sun, and help in conserving soil.

Acknowledgements

We thank Professors Allen M. Solomon and Herman H. Shugart for supporting this work through a summer fellowship to PVD for the Young Scientists Summer Program at the International Institute of Applied Systems Analysis (IIASA), Laxenburg, Austria.

References

Botkin, D.B., J.F. Janak & J.R. Wallis. 1972. Some ecological consequences of a computer model of forest growth. *J. Ecol.* **60**:849-872.

Canham, C.D. 1989. Different responses to gaps among shade-tolerant tree species. *Ecology* **70**:548-550.

Chidumayo, E.N. 1987. Species structure in Zambian miombo woodland. *J. Trop. Ecol.* **3**:109-118.
Chidumayo, E.N. 1988a. Regeneration of *Brachystegia* woodland canopy following felling for tsetse-fly control in Zambia. *Trop. Ecol.* **29**:24-32.
Chidumayo, E.N. 1988b. Estimating fuelwood production and yield in regrowth dry miombo woodland in Zambia. *For. Ecol. Manage.* **24**:59-66.
Chidumayo, E.N. 1990. Above-ground woody biomass structure and productivity in a Zambezian woodland. *For. Ecol. Manage.* **36**:33-46.
Chidumayo, E.N. 1991. Seedling development of the miombo woodland tree *Julbernardia globiflora. J. Veg. Sci.* **2**:21-26.
DeAngelis, D.L., R.L. Gardner & H.H. Shugart. 1981. Productivity of forest ecosystems studied during the IBP: The woodlands dataset. Pp 567-672 in: Reichle, D.E. *Dynamic properties of forest ecosystems.* Cambridge University Press, Cambridge.
Doyle, T.W. 1981. The role of disturbance in the gap dynamics of a montane rain forest: An application of a tropical forest succession model. In: D.C. West, H.H. Shugart & D.B. Botkin (eds) *Forest Succession: Concepts and Application.* Springer-Verlag, New York.
Ernst, W.H.O. 1971. Zur Okologie der Miombo-Walder. *Flora* **160**:317-331.
Ernst, W.H.O. 1988. Seed and seedling ecology of Brachystegia spiciformis, a predominant tree component in miombo woodlands in South Central Africa. *For. Ecol. Manage.* **25**:195-210.
Glover, P.E. 1968. The role of fire and other influences on the savannah habitat, with suggestions for further research. *East Afr. Wildl. J.* **6**:131-137.
Hardcastle, P.D. 1977. *A Preliminary Silvicultural Classification of Malawi.* Forestry Research Record No. 57, Forest Research Institute of Malawi, P.O. Box 270, Zomba, Malawi.
Hopkins, B. 1965. Observations on savanna burning in the Oklokemeji Forest Reserve, Nigeria. *J. Appl. Ecol.* **2**:367-381.
Lawton, R.M. 1964. The ecology of the *Marguesia acuminata* (Gilg.) R.E.fr. evergreen forests and the related chipya vegetation types of north-eastern Rhodesia. *J. Ecol.* **52**:467-479.
Lawton, R.M. 1972. *An ecological study of miombo and chipya woodland with particular reference to Zambia.* Oxford, Ph.D. thesis, 1972.
Lawton, R.M. 1978. A study of the dynamic ecology of Zambian vegetation. *J. Ecol.* **66**:175-198.
Leemans, R. 1989. Description and simulation of stand structure and dynamics in some Swedish forests. Acta Univ. Ups., *Comprehensive Summaries of Uppsala Dissertations from the Faculty of Science* **221**. 44pp. Uppsala.
Leemans, R. 1991. Sensitivity analysis of a forest succession model. *Ecol. Mod.* **53**:247-262.
Leemans, R. & I.C. Prentice. 1989. FORSKA, a general forest succession model. *Meddelanden från Växtbiologiska Institutionen, Uppsala* 1989:**2**.
Leemans, R. & W. Cramer. 1991. *The IIASA database for mean monthly values of temperature, precipitation, and cloudiness of a global terrestrial grid.* WP-90-41, IIASA, Laxenburg.
Malaisse, F. 1978. The miombo ecosystem. Pp 589-606 in: Tropical Forest Ecosystems. *Natural Resources Research* **XIV**, Unesco, Paris.
Meyer, H.A. 1940. A mathematical expression for height curves. *J. For.* **38**:415-420.
Muller, J.M. 1982. *Selected climatic data for a global set of standard stations for vegetation*

science. Junk, The Hague.
Pastor, J. & W.M. Post. 1985. *Development of a linked forest productivity-soil process model.* ORNL/TM-9519, Oak Ridge, Tennessee.
Platt, W.J. & D.R. Strong. 1989. Gaps in forest ecology. *Ecology* **70**:535.
Prentice, I.C. & R. Leemans. 1990. Pattern and process and the dynamics of forest structure: a simulation approach. *J. Ecol.* **78**:340-355.
Reichle, D.E. 1981. *Dynamic properties of forest ecosystems.* Cambridge University Press, Cambridge.
Shugart, H.H. & D.C. West. 1977. Development of an Appalachian deciduous forest successional model and its application to the impact of the chestnut blight. *Journal of Environmental Management* **5**:161-179.
Shugart, H.H. & I.R. Noble. 1981. A computer model of succession and fire response of the high-altitude Eucalyptus forest of the Brindabella Range, Australia Capital Territory. *Aust. J. Ecol.* **6**:149-164.
Shugart, H.H., M.S. Hopkins, I.P. Burgess & A.T. Mortlock. 1981. The development of a succession model for subtropical rain forest and its application to assess the effects of timber harvest at Wiangarree State Forest, New South Wales. *J. Environm. Manage.* **11**:243-265.
Shugart, H.H. 1984. *A Theory of Forest Dynamics.* Springer-Verlag, New York.
Solomon, A.M., D.C. West & J.A. Solomon. 1981. Simulating the role of climate change and species immigration in forest succession. Pp 154-177 in: D.C. West, H.H. Shugart & D.B. Botkin (eds) *Forest Succession: Concepts and Application.* Springer-Verlag, New York.
Stromgaard, P. 1985. Biomass estimation equations for miombo woodland, Zambia. *Agrofor. Syst.* **3**:3-13.
Swaine, M.D. & T.C. Whitmore. 1988. On the definition of ecological species groups in tropical rain forests. *Vegetatio* **75**:81-86.
Trapnell, C.G. 1953. *The soils, vegetation and agricultural systems of North-eastern Rhodesia.* Government Printer, Lusaka, 146 pp.
Trapnell, C.G. 1959. Ecological results of woodland burning experiment in Northern Rhodesia. *J. Ecol.* **47**:129-168.
Van Daalen, J.C. & H.H. Shugart. 1989. OUTENIQUA - A computer model to simulate succession in the mixed evergreen forests of the southern Cape, South Africa. *Landsc. Ecol.* **2(4)**:255-267.
Watt, A.S. 1947. Pattern and process in the plant community. *J. Ecol.* **35**:1-22.
Werger, M.J.A. and B.J. Coetzee. 1978. The Sudano-Zambezian Region. Pp 301-462 in: M.J.A. Werger (Ed) *Biogeography and Ecology of Southern Africa.* The Hague, Junk, 1439 pp. (Monographiae Biologicae Vol. 31).
West, D.C., H.H. Shugart & D.B. Botkin. Eds. 1981. *Forest Succession, Concepts and Application.* Springer-Verlag, New York.
Wetherald, R.T. & S. Manabe. 1986. An investigation of a cloud cover change in response to thermal forcing. *Clim. Change* **8**:5-23.
White, F. 1979. The Guineo-Congolian Region and its relationships to other phytochoria. *Bull. Jard. Bot. Nat. Belg.* **49**:11-55.
White, F. 1983. *The vegetation of Africa.* Unesco, Paris, 356 p.
Whitmore, T.C. 1989. Canopy gaps and two groups of forest trees. *Ecology* **70**:536-538.

Australia's wet-dry tropics: identifying the sensitive zones

R.J. Williams
G.D. Cook
R.W. Braithwaite
A.N. Andersen
L.K. Corbett.
CSIRO Division of Wildlife and Ecology, Tropical Ecosystems Research Centre
PMB 44 Winnellie, N.T. 0821. Australia

Introduction

Global climatic change is a likely consequence of the increasing levels of the "greenhouse gases" in the atmosphere following industrialisation. The predicted increases in atmospheric CO_2 concentration, rising sea levels, and changes in both temperature and rainfall regimes are all likely to affect the world's biota (IPCC, 1990). The aim of this paper is to discuss the possible responses of the biota of the tropical savannas of northern Australia to global atmospheric and climatic changes which are predicted to occur by 2050.

The wet-dry tropics of northern Australia occupy about 30% of the continent (Figure 1). Rainfall in the region is highly seasonal, with a distinct summer maximum - the dry season is prolonged and drought is severe. The major land uses within this area include nature conservation, recreation, tourism, mining and pastoralism. The biota within this region, which has evolved in conjunction with stress and disturbance (Walker & Menaut, 1988), is diverse and has high conservation value. Our discussion centres on the tropical savannas of the northern section of the Northern Territory (12° - 18° S), which corresponds to the Darwin and Gulf, Victoria River and Barkly Tablelands districts of the Northern Territory (Bowman *et al.*, 1988b). Special emphasis will be placed on the humid savannas of the coastal and sub-coastal regions, where rainfall is highest. Our scenarios for climate change are based on the recent CSIRO4 General Circulation Model (GCM) (CSIRO, 1990), and are discussed for a range of scales from the sub-continental to the level of the plant community.

Scenarios for climate change in northern Australia

As yet, the spatial resolution of global models of climate change is relatively low, and different GCMs produce different predictions of temperature and precipitation changes at a regional scale (Henderson-Sellers, 1991). GCMs are therefore of insufficient sensitivity to allow accurate prediction of biotic reponses at a regional scale (Woodward, 1990). Of the current GCMs we have chosen the CSIRO4 model (CSIRO, 1990). Scenarios presented by this model for northern Australia based on a doubling of atmospheric CO_2 by the middle of next century are similar to earlier scenarios of Pittock and Nix (1986) and Pittock (1988). This model predicts a more positive Southern Oscillation Index (SOI). Changes which would result from this include increases in annual rainfall of 10%, and increases in the intensity of rainfall events by 20-30%. Humidity during the wet season may also rise. Temperatures are expected to increase by 1-2°C in northern coastal regions, and by 2-4°C inland. A 2°C increase in mean annual temperature would increase the number of summer days when temperature exceeds 35°C from one to 25 in Darwin, and from 66 to 76 in Tennant Creek (Moffatt, 1991). Sea levels are expected to rise by 20-30 cm. In addition, it is predicted that cyclonic activity will increase, with an extra two to three tropical cyclones forming in the Australasian region each year for every 1°C rise in sea surface temperatures (Holland *et al.*, 1988). Drops in the core pressure within cyclones of 6-24 HPa per 1°C rise in sea surface temperature are possible, hence more intense cyclones, associated with more frequent extreme cyclonic surges (Love, 1988) may eventuate.

Present environment of the northern Northern Territory

To assess the potential responses of savannas to climatic change, it is necessary to review the current understanding of interactions between abiotic and biotic elements of the ecosystems. We therefore present a summary of the known and hypothesised descriptions and determinants of Australian tropical savannas.

Physiography and soils

North of 20° S, the landscape is generally below 300 m a.s.l. Consequently, topography has a negligible effect on the macro-climate. Major landforms include sandstone plateaux and outliers, coastal and subcoastal plains, and extensive flood plains nearer the coast. Extensive, gently undulating plains are common inland. The freshwater flood plains developed within the last 4000 years from marine sediments laid down approximately 7000 years BP (Williams, 1991; Chappell & Woodroffe, 1985). However, most terrestrial landscapes are much older than this with extensive lateritisation common on the dominant paleozoic sediments. Quartz sands cover extensive areas of this region particularly in the south-west (Tanami). Sandstones and siltstones are common in the north-east (Arnhem) and west (Victoria River District, VRD) while low grade metamorphosed sedimentary rocks occur in the north-west. Areas of limestone are scattered throughout the region and basic igneous rocks occur in the VRD. Clays cover extensive areas of the south-east on the Barkly Tableland.

Generally the parent materials for soil formation are impoverished. The landscape is old, and at a late stage of levelling, so that weathering fronts are relatively deep and nutrient release from parent materials is of minor importance to the nutrient status of soils. Palaeosols are common (Mott *et al.*, 1985); hence the distribution of soils can not be understood from the current

Figure 1.
Map of Northern Territory with rainfall isohyets (mm).

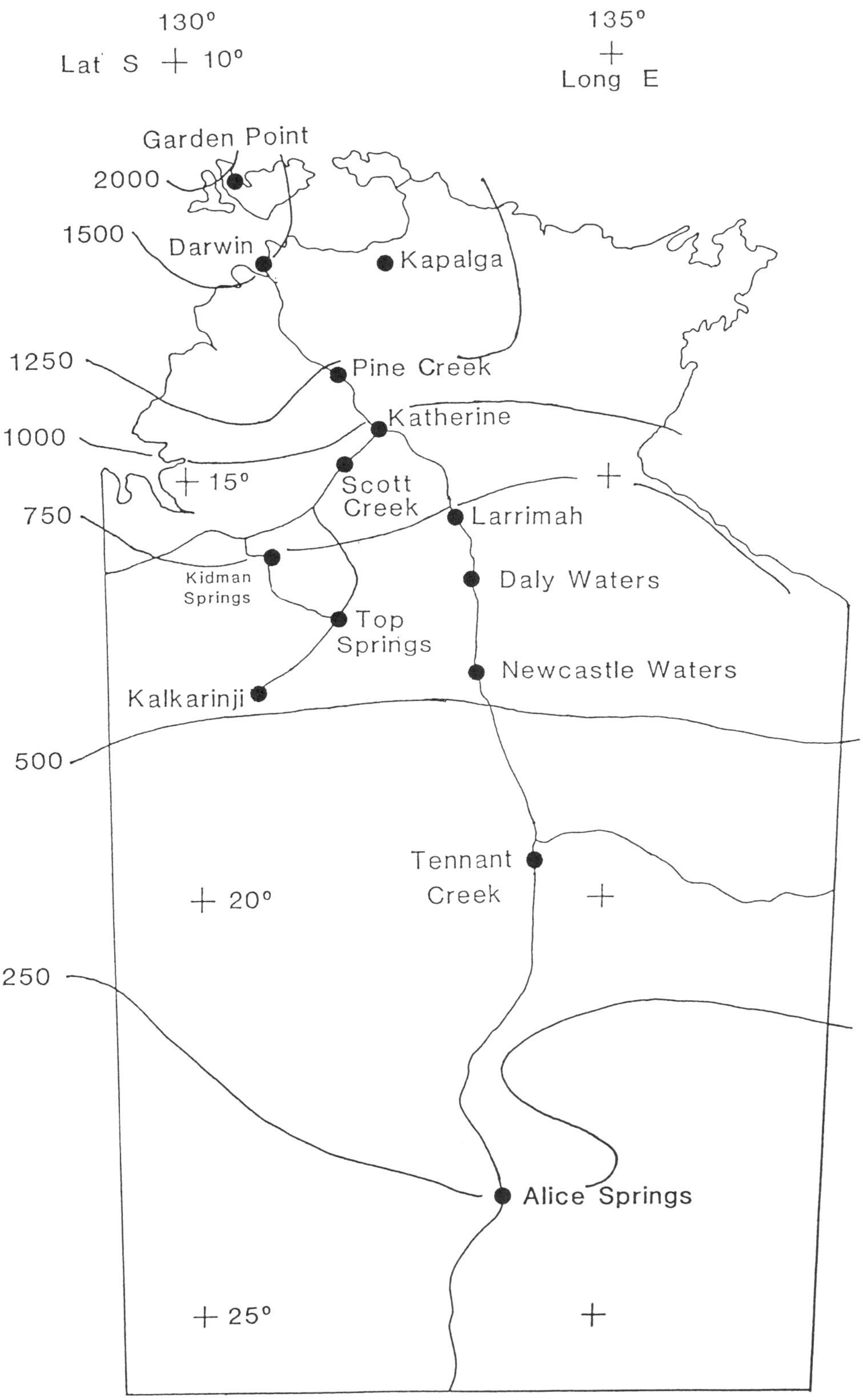

climate. Soils are usually highly leached and of coarse texture, the most common groups being skeletal soils and the sands (lithosols, entisols, regosols, arenosols) and (massive) earths (yermosols, ferralsols, acrisols, nitosols). Duplex soils are locally important in the east and north-west. Cracking clays (vertisols) are the most fertile soils of the region and occur mainly in the south-east (Barkly Tableland) where rainfall is low, and on the coastal floodplains where they are subject to flooding with fresh water each wet season. The distribution of soil types is patchy, with the mosaic diverse at a scale of 1:1,000,000 (Northcote, 1968). Nutrient contents are generally low; total N = 0.08-0.19%, available P 5-59 ppm; CEC = 8-45 meq % (Mott *et al.*, 1985).

Climate

The present climate is wet-dry tropical in the north to dry tropical in the south, with a strong north-south gradient in plant available moisture. Annual rainfall decreases with increasing latitude, from about 2000 mm at Melville Island (11°25' S; 130°25'E) to 434 mm at Tennant Creek (19°38' S; 143°11' E) a decrease of more than 1 mm per annum per kilometer. In this region, the rainfall is distinctly seasonal, with over 90% falling in summer (October - April). Thus the greatest predictable annual event is that there will be drought between May and September, during which time the upper 1.5 m of soil is below permanent wilting point (Bowman & McDonough, 1991). There is, however, considerable spatial and temporal variation in both the onset of the wet season and the amount of rain within the wet season (Taylor & Tulloch, 1985). Annual rainfall for Darwin varies between 800 and 2600 mm p.a., and there may be 4-6 weeks difference in the timing of the onset of the wet season from year to year. Rain is due to convective storms, tropical cyclones and monsoonal troughs. Variation in the latter is responsible for most of rainfall variability within wet seasons. Between onset and retreat, the monsoon has active and inactive periods, with a mean return period of active bursts of approximately 40 ± 10 days (Holland, 1986).

Temperature is megathermal and equable in northern coastal areas; mean monthly maxima at Darwin range from 30-35°C and the corresponding minima are 18-27°C. Light and photoperiod are both relatively constant between seasons and years (Ridpath, 1985). With increasing latitude, differences in seasonal photoperiod are amplified, and variation in both diurnal and seasonal temperature regimes is greater.

The climate of Australian savannas has varied considerably over the Quaternary, with some evidence that aridity has increased over the past 10,000 years (Mott *et al.*, 1985). The climate at present is drier than that of 7000-9000 BP (Williams, 1991) and there is some evidence that the last century has been drier than the previous 900 years (MAJ Williams, pers. comm.; J Schulmeister, PhD thesis Australian National University). The vegetation is therefore in a state of constant ecological change (Mott *et al.*, 1985) and current vegetation may be responding to abiotic events of both the recent past and the present.

Biota

The plant and animal assemblages are both diverse and patchy. The region as a whole has suffered less than central and southern Australia from wide-spread, high impact exogenous disturbances (Woinarski & Braithwaite, 1990). Hence extinctions have been rare, and the area as a whole has high conservation value. The vegetation varies in structure from closed forest toopen grassland, but is mostly savanna forest and woodland, with an overstorey of trees or tall shrubs, and grassy understorey. Grassland occurs on the heavier soils of the floodplains and Barkly Tableland. Closed "monsoon" forests and thickets occur where available soil water is high both within and between years, such as at the bases of escarpments and in gullies (Russell-Smith, 1991).

The savannas are dominated by *Eucalyptus* species, but include both sclerophyllous and broad leaf trees. Deciduous tree species are a feature, constituting some 25% of the tree/tall shrub taxa in both forests and woodlands of the Alligator Rivers region (Taylor & Dunlop, 1985). Obligate deciduous genera include *Terminalia, Brachychiton, Cochlospermum* and *Ficus*. Some of the eucalypts are deciduous or semi-deciduous. Understorey grasses include both tall (2-3 m) annuals (eg. *Sorghum*) and perennials (*Heteropogon*), and numerous shorter grasses, both perennial and annual. Other common annual genera include *Alloteropsis, Setaria* and *Pseudopogonantherum*; other perennial genera are *Chrysopogon, Themeda, Dichanthium, Plechtrachne* and *Triodia*. The majority of the dominant grasses belong to the tribe Andopogonae. All of the dominant grass species have the C_4 photosynthetic pathway (Hattersley, 1983), whereas the woody species are all C_3. Understorey forbs and small shrubs are diverse, and most do not retain their foliage over the dry season (Taylor & Dunlop, 1985).

The structure and composition of the savanna vegetation varies primarily in relation to moisture supply and nutrients (Walker & Menaut, 1988). Secondary determinants are fire and herbivory. At the sub-continental scale (1000 km), between Darwin and Tennant Creek, both the height and projective foliage cover of the trees decrease linearly with decreasing rainfall (Figure 2). In the north, tall annual grasses such as *Sorghum intrans* dominate the ground stratum, but the cover and diversity of annual grasses both decrease with increasing latitude (Table 1).

Table 1.
Index of abundance of annual grasses in dominant savanna vegetation units (of Wilson *et al.* 1990) between Darwin and Tennant Creek. Latitudinal range of vegetation units indicated. Index calculated as: [(mean proportional cover of annual species within unit) x (number of annual species/total grass species in unit) x 100].

Unit	Latitude (°S)	Index
4	12 - 13	13
15	12 - 15	7
21	13 - 14.5	5
13	14 - 15.5	2
17	14.5 - 16.5	3
20	15.3 - 17	1
25	14.5 - 18	2
26	16.3 - 18	6
39	17 - 19	2
36	17.5 - 19.7	1
42	18.5 - 19.4	1

Figure 2.
Relationship between tree height (a) and projected foliage cover (b) and rainfall for various eucalypt savannas between Darwin and Tennant Creek. Source: Wilson *et al.* (1990).

(a)

(b)

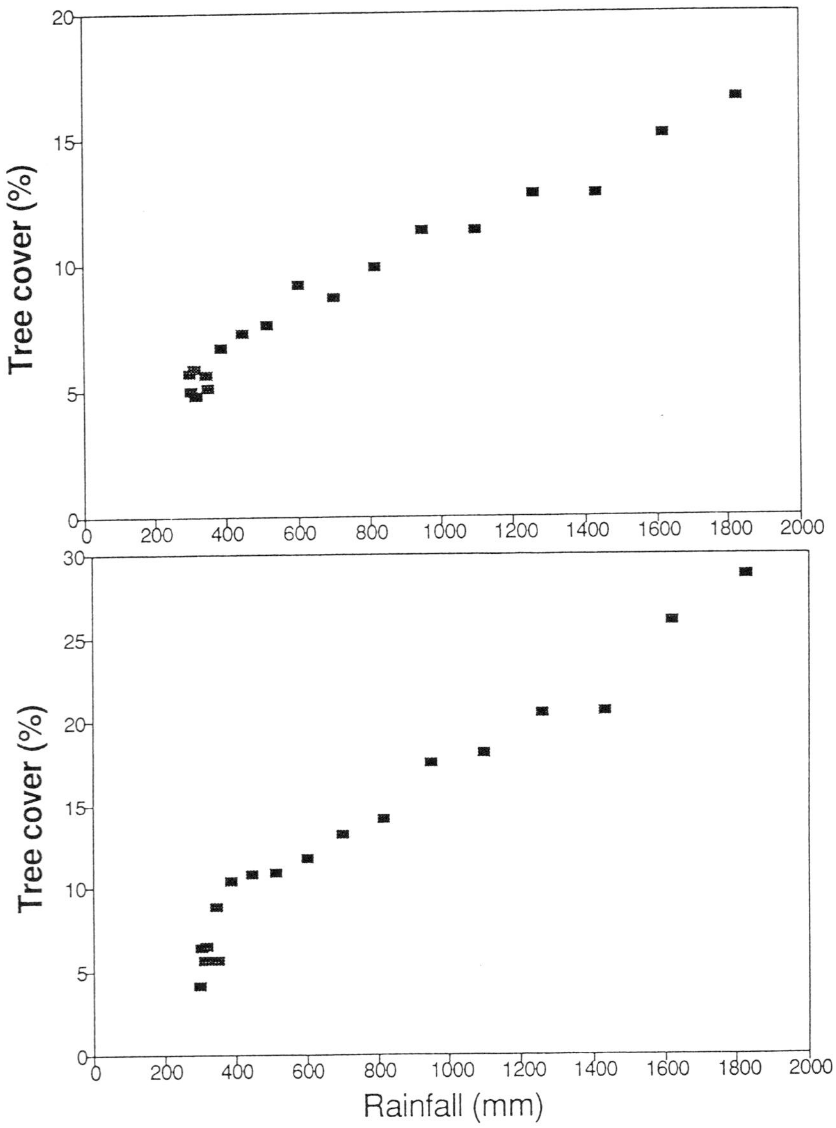

At the river catchment, or savanna-to-wetland, scale (10 km), in the humid coastal and subcoastal regions of the Darwin and Gulf region, vegetation varies from mangrove forests at the edge of the major, perennial rivers and creeks, to grassland and sedgeland on the clay soils of the floodplains, with eucalypt forest and woodland on the coarser, well drained soils of the ridges (Figure 3). Margin communities include monsoon forest, paperbark forest and pandanus thickets. The 12,700 km of the more coastal Stages I and II of Kakadu National Park (12° S; 132° E) consists of mesic forest (10%), wetlands (13%), sparse woodlands of the sandstone escarpents, and forests (24%) and woodlands (33%) of the plains (Braithwaite & Estbergs, 1985). Within this region the rivers have tidal ranges of 3 m up to 50 km inland.

Figure 3.
Idealised topographic-soil-vegetation relationships for savanna, margin and floodplain communities for the Alligator Rivers catchments in Kakadu National Park.

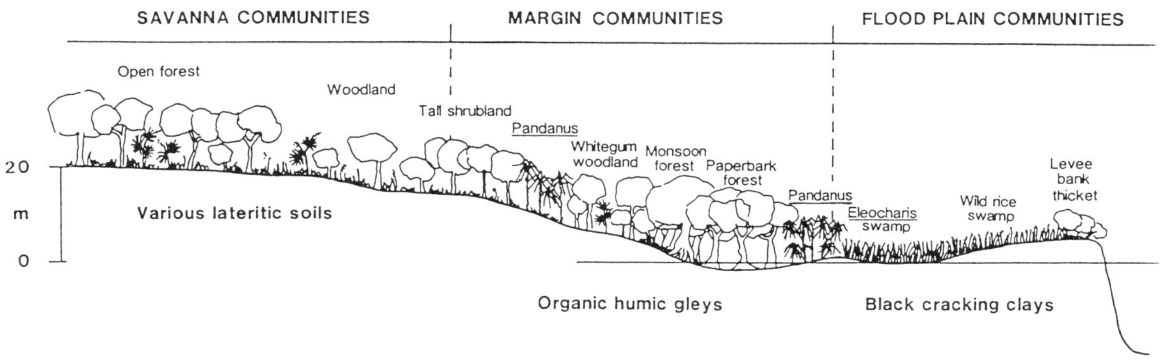

At the scale of sub-catchments, or ridge-to-seep within humid savannas, (1 km) creeks feeding the major rivers are ephemeral. Soil water availability decreases with increasing distance from these streams and other seepage zones. Forest savannas occur on the deeper, loamy soils near the low ridges, with woodland savannas on the shallower, seasonally flooded soils near the seeps. The cover of annual grasses is similar in both zones, but the cover of perennial grasses in the seep zones is twice that of the ridge zones (Table 2). Both alpha and beta diversity of the understorey are high in each habitat, but the former tends to increase towards the ridges and the latter is greatest near the seeps (R.J. Williams & R.W. Braithwaite, unpublished data). At this scale both canopy cover and height of trees are inversely related to both soil rockiness and degree of waterlogging (G. Cook, unpublished data). There also appears to be an inverse relationship between projective foliage cover of the woody canopy and that of the grassy understorey (Bowman 1986; Figure 4).

Table 2.
Average cover (%) of annual and perennial grasses in the ground stratum of forest on well-drained lateritic soils, and woodland on seasonally waterlogged soils of seepage zones at Kapalga, Kakadu National Park.

Forest		Woodland	
Annual	Perennial	Annual	Perennial
76	47	71	95

Figure 4.
Relationship between basal area of trees and projected standing crop of grass for 21 forest and woodland savanna sites at Kapalga, Kakadu National Park.

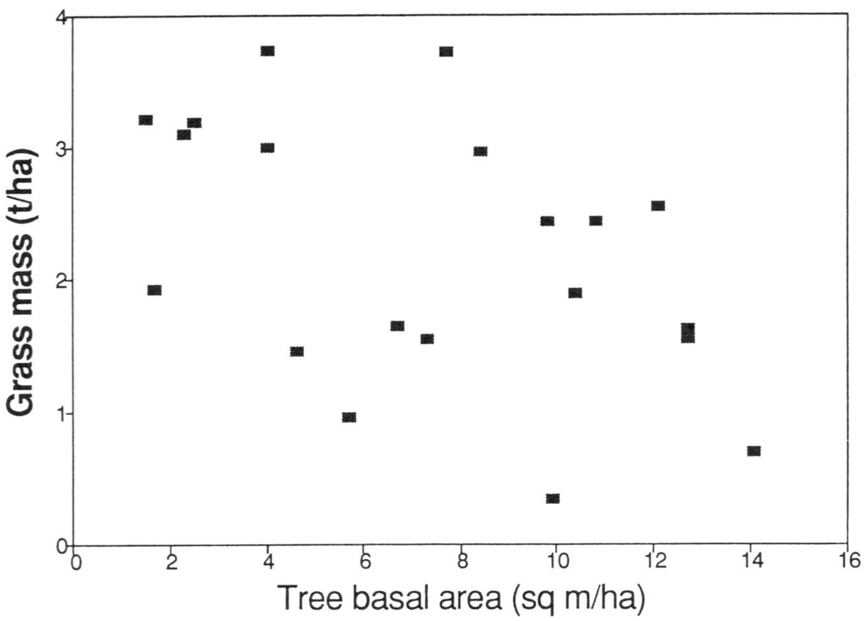

Savanna dynamics

The seasonal patterns of plant growth differ between trees and the understorey. Most non-deciduous woody species grow in the dry season (Fensham, 1990a), as demonstrated by Sarmiento and Monasterio (1983) for American savannas. Growth of woody sprouts and saplings after fire occurs predominantly in the current dry season, followed by a dormant period in the first wet season following fire. Leaf flush in adult trees may occur throughout the dry season, with distinct peaks in the wet season (Figure 5). In the obligate deciduous species, leaf flush occurs in the 4-6 weeks prior to the onset of the wet season (R.J. Williams, unpublished data). Total leaf fall peaks late in the dry season (Table 3). In contrast more than 90% of herbaceous biomass is produced in the wet season (Mott *et al.*, 1985). Productivity of the understorey is between 100-300 g m^{-2} yr^{-1} in coastal/subcoastal forests (Mott *et al.*, 1985). There are few data on productivity in trees. Regeneration is primarily via vegetative means in the woody species. The capacity to resprout vegetatively following disturbance is widespread, and seedling recruitment in trees is uncommon (Werner, 1986).

Table 3.
Seasonal patterns of leaf fall from trees (g m^{-2} d^{-1}; ± standard deviation) in humid eucaltypt forest and woodland savannas at Kapalga, Kakadu National Park. G. Cook, unpublished data.

	Nov/Dec	Jan/Feb	Mar/Apr	May/Jun	Jul/Aug	Sep/Oct
Mean	1.25	0.51	0.86	1.48	2.61	1.33
SD	0.09	0.10	0.19	0.22	1.06	0.32

Figure 5.
Relationship between leaf flush in trees (>3 m height), herbivory on those trees, and rainfall (mm) over two successive seasons for a savanna forest at Kapalga, Kakadu National Park. Leaf flush index calculated as extent of flush (3- point ordinal scale) for 40 species at 24 sites. Herbivory index calculated in a similar manner to leaf flush index, and corrected for leaf biomass (R.W. Braithwaite, unpublished data).

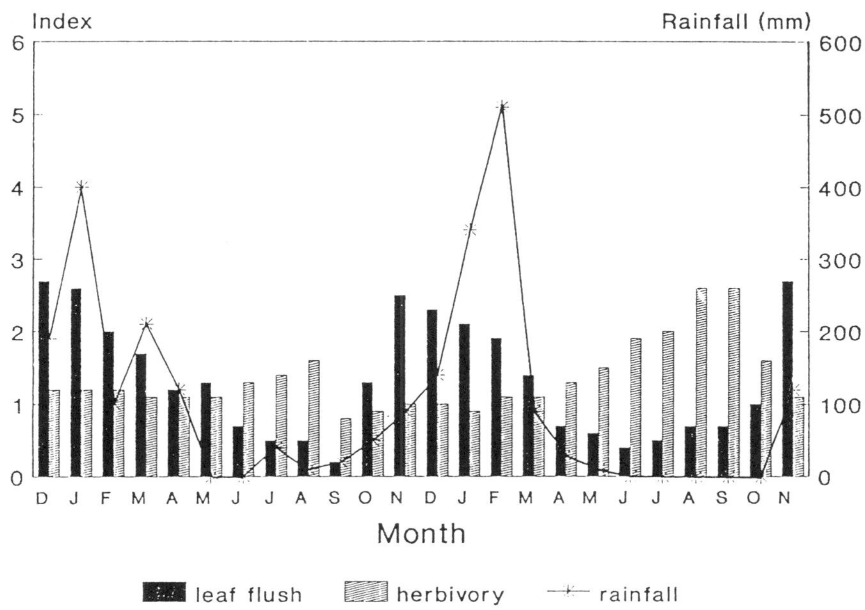

The ecology of animal assemblages has been studied at all three scales, but is most intensive at the catchment and sub-catchment scales (Braithwaite, 1990). At the landscape scale, the diversity of vertebates is high, especially in the coastal and subcoastal regions, which reflects mosaic of landscape types (Woinarski & Braithwaite, 1991). The Kakadu region as a whole contains refugia for many rare and threatened mammals. At catchment scales the diversity of small mammals is greater in savannas than wetlands, whilst within savannas, it is greatest near the seeps (Braithwaite, 1990). Some honeyeaters, raptors and mammalian predators move along extensive gradients to exploit the different seasonal resources (Braithwaite & Werner, 1987). The life history strategies of the vertebrate fauna are geared to exploiting the changing resources available throughout the annual monsoon cycle, and coping with the annual drought. For example, different lizard species breed at different times of the year (James & Shine, 1985). Magpie geese migrate from wetland to wetland as these dry out differentially, thereby presenting *Eleocharis* corms in the mud at different stages of the dry season (Morton & Brennan, 1991).

Herbivory, almost all of which is by invertebrates, is variable along gradients at these scales (Andersen & Lonsdale, 1990). However, at the sub-catchment scale herbivory is greater during years of higher rainfall, especially in the dry season subsequent to wet seasons when rainfall is relatively high (R.W. Braithwaite, unpublished data, Figure 5).

Fire is an integral part of the system, and the biota has evolved in conjunction with fire. The vegetation of northern Australia is particularly fire prone, due to the high productivity of the ground stratum during the wet season, the extended dry season during which fuels cure, and the dry electrical storms in build-up to the wet. In pre-aboriginal times, fires occurred in the build-up to the wet season, when fuels were cured, lightning frequent, but rain uncommon. Aboriginal man burnt the savannas for 9 months of the year, beginning at the end of the wet season. At present fires occur annually-biennially (Braithwaite & Estbergs, 1985; Braithwaite, 1991). Pre-aboriginal, aboriginal and European regimes have distinctly different central tendencies, but overlap broadly. Fires in the early dry season are generally less intense and more patchy than those of the late dry season (Braithwaite & Estbergs, 1985). Over 50% of all lowland forests carry fire each year (Press, 1988). There is considerable year to year variation, however, in the intensity of fires within a given season (A.M. Gill, unpublished data). The timing of fires may have minimal effects on overall floristic composition of the vegetation, as effects of fire appear to depend substantially on site factors (Bowman *et al.,* 1988a). However, occasional intense fires may cause considerable changes in the composition of dominant tree species due to differential mortalities between species to such fires (Lonsdale & Braithwaite, 1991). In the absence of fire, the woody component of the vegetation becomes more abundant, and grass cover, especially that of annuals, decreases (Bowman *et al.,* 1988a).

A common faunal response to fires, that of habitat selection of the successional seres followingfires many years apart, is not possible in ecosystems with such a high frequency of fire. Instead, mobile species such as birds and larger mammals exploit burning or burnt patches as their habitat. Hunting during fires, scavenging after fires, foraging on resprouting vegetation and preying on animals drawn to these conditions are all short term opportunities used by more motile species of animals (Braithwaite & Estbergs, 1987). Other species utilise the habitat created by fires of different types. Here, the variation in both phenological condition of the vegetation and intensity of fire produce a range of habitat types which are favoured by various animal species (Kerle, 1985; Braithwaite, 1987; Woinarski, 1990).

Some considerations of biotic response to climate change

Species may respond to climate change in behavioural, phenological and evolutionary ways, and by altered rates of growth, reproduction and migration. Two aspects of climatic change must be considered, namely; changes in mean conditions, and changes in the extremes associated with average conditions. Community attributes such as productivity are likely to change in response to changes in mean conditions (eg. Pittock & Nix, 1986). However, the variance associated with such changes in mean values will have substantial impacts on both productivity, and compositional attributes of community dynamics, as the latter appear to be determined principally by stochastic processes such as recruitment and mortality (Ojima *et al.,* 1991). Given that the present GCM scenarios cannot predict the occurrence of extreme events, and that their role in community organisation, although substantial, is little understood, then the task of developing predictive models of biotic response to climate change is extremely difficult. The role of migration is especially important, and its role in response to greenhouse related climate change is likely to be far greater than that of evolution (Bradshaw & McNielly, 1991; Huntley, 1991) as the rate of such change is likely to be more rapid than for any other period in past 160,000 years (IPCC, 1990). Animals are likely to be more responsive than plants to climate change, as they are more motile. Insects in particular are likely to be highly responsive, due to their high sensitivity to climatic variables, high reproduction rates and short generation times. With respect to trees which are relatively slow-growing and long-lived, the potential rates of migration are likely to be much less than the predicted rate of climatic change (IPCC, 1990). Expansions and contractions of range are likely to occur at different rates, the former will be affected by events with a periodicity of around 15 years, the latter by those with periodicities greater than 150 years (Miles, 1979; Woodward, 1987).

There may be, however, strong barriers to potential migration in response to climate change. For example, rainfall and temperature "belts" may shift latitude, but onto completely different soil associations. In addition, many landscapes have been altered by human activity, which may create extra barriers to movement (IPCC, 1990). Disruption of plant-animal interactions (pollination, dispersal, herbivory) is also a secondary force driving the responses of vegetation to climate change. In all cases, species with specialised and restricted habitats, or specialised diets and reproductive processes may decline drastically in numbers. Hence, changes in community composition and structure are likely to be identified sooner in transitional zones between vegetation types than elsewhere (IPCC, 1990).

Climatic change and ecosystem response: issues and caveats for Australian savannas

The predicted changes in the temperature and rainfall regimes of the tropical life-zone belts are smaller than those for temperate and high latitude areas (IPCC, 1990). The predicted rise in rainfall is also small when considered in the context of present patterns of temporal and spatial heterogeneity of rainfall. An increase in annual rainfall of 10% is well within the normal range of coefficients of variation between years, which range from 20% at Darwin to 40% at Tennant Creek. Depending on the site, between 70 years and 300 years of further records would be required to detect an increase in mean annual rainfall of 10% (Table 4). Because rainfall is one of the principal abiotic determinants of savanna biology, it is therefore quite possible that there will be little or no detectable response in the biotic factors, especially the vegetation within the next century. In his model studies, Woodward (1991) estimated that 50 years of environmental change will be needed before significant changes in diversity are likely to occur.

Table 4.
Rainfall patterns at a range of sites between Darwin and Tennant Creek. Sites - Darwin, Pine Creek, Larrimah, Daly Waters, Newcastle Waters and Tennant Creek. For each site the data in each column are latitude (°S); mean annual rainfall (mm); coefficient of variation (%); the number of yearly records (n) and the number of years needed to detect a 10% increase in mean annual rainfall (D10%; at t=0.95) given current coefficients of variation.

Site	Dwn	Pn Ck	Lar	Dal W	New W	Te Ck
Lat S	12.25	13.49	15.35	16.16	17.23	19.38
Mean	1660	1140	800	665	490	450
CV%	21	21	31	31	41	40
n	47	95	33	96	94	19
D10% yr	73	71	170	145	268	299

Changes in mean conditions may also affect the occurrence of extreme events. For example, if mean annual rainfall in Pine Creek (13°49' S) were to increase by 10%, but without a change to the shape of the distribution of annual rainfall between years, substantial changes in the frequency of occurrence of wet and dry years would eventuate. Infrequent wet years when rainfall is 1650 mm, and which occur approximately every 60 years, would occur every 20 years (Figure 6). Conversely, dry years with periodicities of about a decade will have return times of several decades.

With respect to the dominant grasses, it is likely that responses to climate change would be driven more by patterning within seasons than total amounts between seasons. This is especially so for the productivity of the annual grasses of the ground layer as they are affected strongly by stochastic events. The standing crop of annual *Sorghum* may depend on amount of rain prior to floral initiation (Andrew & Mott, 1983; Cook & Andrew, 1991). Early rainfall determines the density of seedlings which establish and survive. After the beginning of December, virtually all plants survive. Presumably very early rains which are enough to allow germination may be followed by dry periods under which mortality is high. Seed productivity per plant, however, is influenced by the density of plants. Increased total annual rainfall and increased late wet season rainfall is likely to affect fuel phenology, by delaying curing of perennial herbaceous vegetation (curing of annual grasses appears to be independent of late rains). Thus variations in the distribution of rainfall within the wet season will have implications for productivity of annual and perennial grasses, and consequently the fire regime, as the grasses, especially the tall annual grasses, are the major fuels in the humid savannas.

The predicted increases in CO_2 are of direct importance to all vegetation. There has been considerable research aimed at elucidating the response of biota to elevated CO_2, but little has been done in the tropics, and even less at the ecosystem scale (Eamus & Jarvis, 1987). There is general agreement that a doubling of the atmospheric CO_2 concentrations would have a fertiliser effect on the vegetation resulting in increases to productivity (Rosenzweig & Dickinson, 1986; Eamus & Jarvis, 1987; Gifford, 1988a,b). Average increases, however, will not necessarily be in direct proportion to increases in CO_2 concentration; an increase of 20-40% with a doubling of CO_2 seems most likely, with the possibility of no change and even reduced productivity on some plants (Gifford 1988b). In addition, in nutrient-poor systems, boosts to productivity may ultimately be limited by the low nutrient capital (Billings *et al.*, 1983).

The effect of CO_2 fertilisation would be greater in C_3 plants than in C_4 plants. This has important implications for savannas as the woody plants are C_3 and grasses primarily C_4. Thus, the trees and tall shrubs may be favoured over the grasses, which may result in greater canopy leafiness, more shade and more dicotyledonous litter on the ground. This may cause grass cover especially that of annual grasses like *Sorghum* to decrease. Such species appear to decline in cover if canopy cover increases and/or thick leaf litter accumulates on ground, as occurs in long term (> 10 years) absence of fire (Fensham, 1990b). Fire intensity might therefore decrease and this would further increase woodiness. This scenario is more likely in forests, where canopy cover is relatively high initially, than in more open communities such as woodlands. In general, however, there is little information on C_3/C_4 interactions in the tropics (Hall & Scurlock, 1991).

Figure 6.
Effect of an increase in mean annual rainfall of 10%, but without alteration of the shape of the distribution of rainfall between years, on the occurrence of infrequent rainfall events.
Years when total annual rainfall is 1650 mm (arrowed) occur approximately once in 60 years (a). Were annual rainfall to increase by 10%, such events would occur once every 20 years (b).

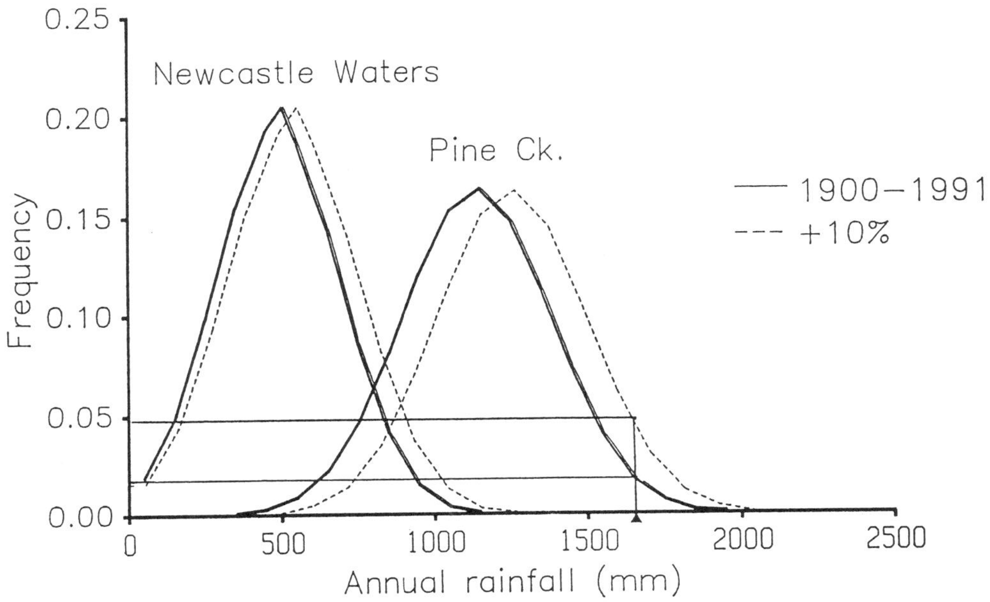

Increased CO_2 concentrations would also increase the water use efficiency of the vegetation, by decreasing stomatal conductance (Eamus, 1991). Water use efficiency may increase at the individual plant level, but at the ecosystem level total water use may not necessarily decrease, due to increased leaf area, and increased evaporation directly from soil as a consequence oflower relative humidity associated with lower transpiration (Gifford, 1988b).

Possible changes at the three scales of observation

Sub-continental

Various sensitivity studies have predicted changes in the vegetation of northern Australia in response to increased rainfall. Pittock and Nix (1986) predict a general shift southwards of the productivity belts under the general scenario of increased rainfall. Given the predictable relationship between both tree height and projected foliage cover and rainfall (Figure 2), it is likely that these parameters will increase with increasing rainfall. In addition, northern Australia as a whole is predicted to be a zone of relatively high sensitivity to changes in plant diversity at the family level (Woodward, 1991), with diversity expected to decrease under a scenario of mean annual temperature rising by 3°C and precipitation increasing by 10%.

The issue of detecting trends in both temporal and spatial patterns of rainfall has been addressed in recent analyses of long-term annual rainfall records for a range of sites in the wet-dry tropics of northern Australia for example, Russell (1981, Darwin), Taylor and Tulloch (1986, Darwin), Mollah *et al.* (1991, Katherine) and Gregory (1991, tropical Queensland). These studies have concluded that variability between years is relatively high (CV 25%- 40%), and, despite the occurrence of sustained periods (decades) of both above and below average annual rainfall, there has been no statistically detectable trend in the amount of annual rainfall over the past century. Thus, despite detectable increases in CO_2 and global temperature over the past 50-100 years (IPCC, 1990) there has been no detectable trends in rainfall patterns inthe northern Australia area.

A 10% change in rainfall is small compared with the rainfall range over which many dominant species and plant commumities presently occur. A number of common taxa have wide latitudinal distributions, and therefore occur over a wide rainfall gradient (Figure 7). For example, on lateritic loams from Melville Island (2,000 mm pa) to near Katherine (980 mm) forests and woodlands of *Eucalyptus tetrodonta* and *E. miniata*, in association with woody species such as *Planchonia, Buchanania, Erythrophleum,* with an understorey of both annual *Sorghum* and perennial grasses, is the dominant plant commumity (Dunlop & Bowman, 1986; Wilson *et al.*, 1990). Over a similar latitudinal range, but on duplex soils of metamorphic origin, the dominant community is an association of *Eucalyptus tectifica* and *E. latifolia*, with a *Sorghum* understorey (Wilson *et al.*, 1990). A 10% increase in the mean rainfall at any point in the range is well within the range of current rainfall conditions over which both of these plant communities occur. Thus, there may be little overall change in the composition and structure of these communities.

One response in the biota is likely to be migration. The migratory potential of the dominant woody plants may be curtailed because at present regeneration by seedlings is rare. Seedling regeneration may, however, assume a greater role under conditions of rapid climate change. Animals, especially insects, are likely to be more motile than plants. For successful migration to occur, the barriers to movement should be few. Effective barriers are likely to occur where expanses of two different soil types abutt one another, for example, where the clays of the Barkly Tableland meet the loams and sands of the Darwin and Gulf region. Alternatively, one extensive assemblage of plants adjacent to another, of substantially different composition, mayalso act as a barrier.

We hypothesised that major landscape boundaries may exist where extensive patches of heterogenous vegetation abutts substantial areas of homogenous vegetation in the direction of the predicted climate change. We therefore analysed the patterns of diversity of both soil and vegetation communities in 0.5 x 0.5 degree blocks along 6 meridians of longitude (130-135° E) over a wide latitudinal range from Melville Island (11° N) to south of Tennant Creek (19° N). We used maps of both the soil (Northcote, 1968) and vegetation (Wilson *et al.*, 1990) at a scale of 1:1,000,000 as our data source. Within each block we listed the communities or soil types (as mapped), and calculated the degree of similarity between adjacent blocks using the dissimilarity (D) measure $D = [1 - (2j/a+b)]$ - where a = number of taxa in block a, b = number of taxa in block b, and j = number of taxa common to block a and block b (Moore & Chapman, 1986). This is a measure of heterogeneity between land units along a gradient, or ß-diversity.

Figure 7.
The distributions of nine common tree species associated with *Eucalyptus tetrodonta* and *Eucalyptus miniata* open forest species. Source: Wilson *et al.* (1990); R. Fensham, unpublished data.

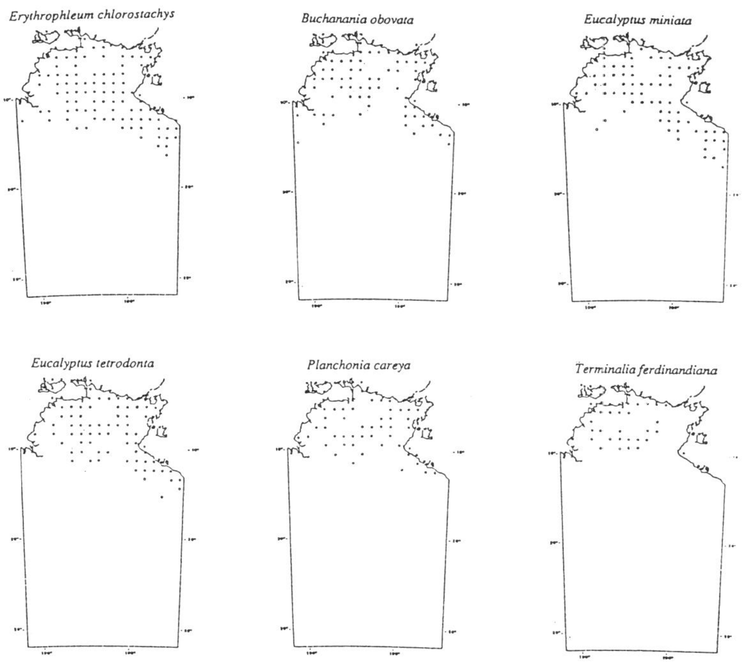

We plotted dissimilarity against latitude, along this 1,200 km transect. Figure 8a shows the pattern along meridian 133° E. From the survey of plant communities, we identified several areas where an area of high heterogeneity abutted one of homogeneity to the south. These correspond to troughs in the graph, and occurred at approximately 13°, 16° and 18° S (Figure 8b). Geographically, 13° represents the southern extent of coastal and subcoastal savanna vegetation; between 16° and 17° there is an extensive and distinctive belt of *Acacia shirleyi* (lancewood) within an otherwise eucalypt-dominated landscape, and 18° is the northern limit of major arid and semi-arid land systems of the Tanami Desert to the west and Barkly Tablelands to the east. The 18° zone corresponds to a major biogeographic boundary proposed by numerous authors, both botanists and zoologists working with various taxa at the sub-continental and continental scale (Spencer, 1896; Burbidge, 1960; Archer & Fox, 1984; Gill *et al.*, 1985; Hnatiuk & Pedley, 1985; Bowman *et al.*, 1988b). It corresponds approximately to the 500 mm rainfall isohyet, and essentially separates the biota of the savanna areas of higher, summer rainfall from that of the arid regions with lower, more sporadic rainfall.

Either the southern or the northern limits to the distribution of many vertebrate species coincides with one or other of these lines. Considering current estimates of species boundaries for birds, mammals, reptiles and amphibians, (Strahan, 1983; Cogger, 1988) there are 12 southern and three northern boundaries that approximate the 13° zone of dissimilarity. The comparative figures for the 16° zone are 67 southern and 8 northern, and for the 18° zone are 10 southern and 16 northern respectively. Thus, the southern limit of many of the tropical vertebrates is at approximately 16°S, whereas the northern limit of many arid zone species is about 18°S.

Figure 8.
Index of between site heterogeneity at the level of the plant community as a function of latitude for plant communities between the Darwin region (11.5° S) and Tennant Creek region (20° S). Source of data: Wilson *et al.* (1990).
(a) The mean pattern with latitude between meridians 130° E and 136°E.

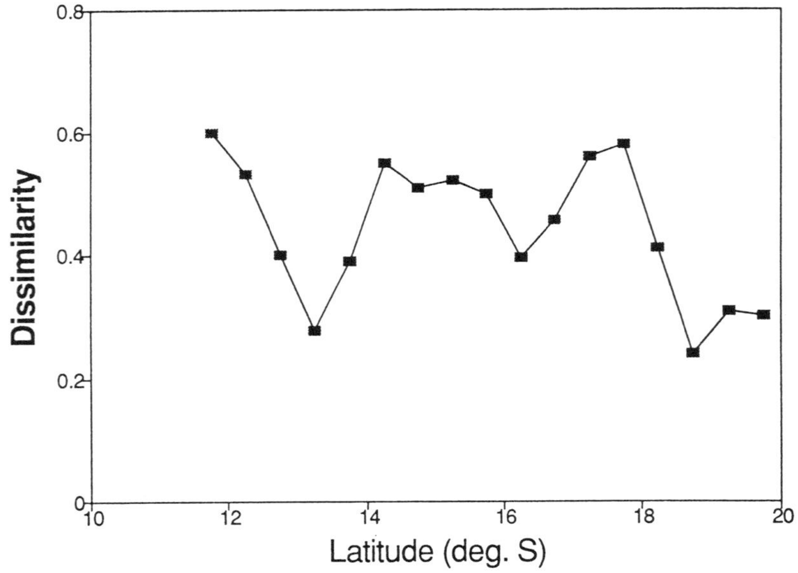

(b) The spatial distribution of the zones of rapid change across the Northern Territory.

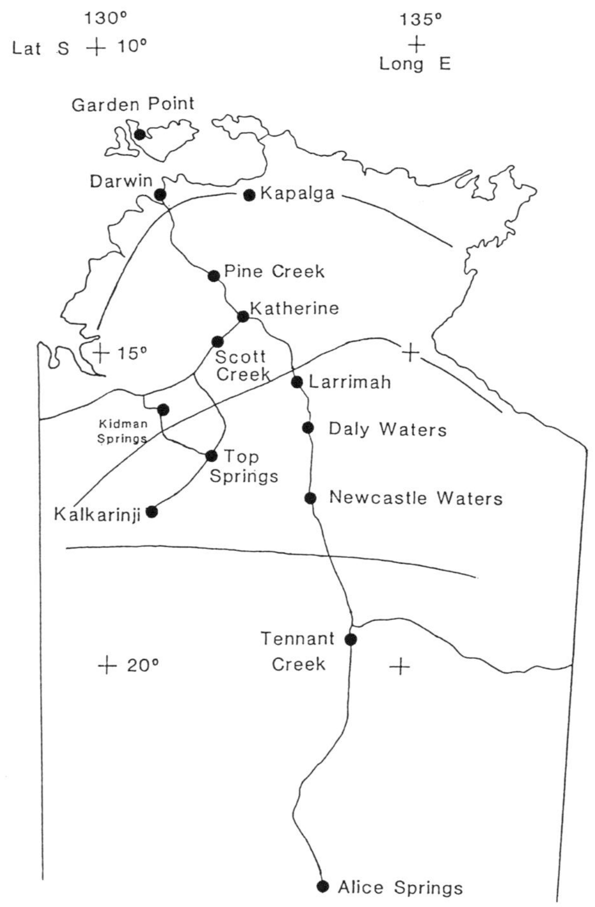

In general, an increase in total rainfall should shift distributional limits of species further south. In the case of tropical species the broad bands of homogeneous vegetation at 16° and 18° may not allow the creation of suitable habitat further south, due to soil type limitations. Consequently, if the northern limit of such a species was also pushed south, the new range would be diminished in size.

On the other hand, in the arid zone the northern distributional limit would be more readily pushed south. However, the northern parts of the arid zone are the least degraded. Consequently, climate change would push species out of the least graded parts of their range. Some species of mammals have already contracted to the northern parts of their range and the predicted climate change would push them out of the best remaining habitat and may thereby push them to extinction. For example, the bilby *Macrotis lagotis* is a rabbit-sized omnivorous marsupial which has been lost from 50-90 percent of its former range in arid Australia and is extinct in the southern part of that range (Kennedy, 1991) where the impact of rabbits, foxes, sheep and cattle has been greatest. The predicted change might result in the Bilby being crushed between the results of climate change and environmental degradation. Other arid zone species which might be vulnerable to this include the rufous hare-wallaby and spectacled hare-wallaby.

A similar interaction between climate change and land degradation may also affect some vulnerable vertebrate species whose distribution is principally tropical. For example, the distributions of both the golden bandicoot *Isoodon auratus* and the golden-backed tree-rat *Mesembriomys macrurus* have contracted northward to a fraction of their former range in European times (Kennedy, 1991), and have disappeared from the cattle country. These species might also be put to great risk by the predicted climate change.

Simulations of climate change at this scale concerning the distribution of the antilopine kangaroo *Macropus antilopinus* were made by Busby (1988) using BIOCLIM, a computer simulation species-climate model. At present *M. antilopinus* is relatively widespread over the top end of the Northern Territory and other sites in N. Australia. On the basis of a 2°C rise in temperature and a 20% increase in rainfall, BIOCLIM predicted that the potential range of the antilopine kangaroo would contract massively. Suitable sites will disappear from the NT, with suitable sites for its existence remaining only in north east Queensland.

Catchment and sub-catchment scales

These are the most practical scales for both sensitivity studies and monitoring of potential responses to climate change, because the most detailed work has been done at this scale. Within ecosystems at this scale the patches and species will vary in their sensitivity to climate change.

An increase in rainfall of 10% will affect rates of soil movement rather than rates of soil formation via weathering. The latter is unlikely to change in the short term (less than a century). Changes to rates of erosion and deposition will be dependent on landscape position. Within the humid savannas, increased erosion is likely to occur where shallow lateritic exposures coincide with minor changes in slope (Figure 9). Run-on zones may experience increased deposition, for example seepage zones associated with ephemeral streams, floodplains, and sandsheets within ravines in the escarpment areas, which are dominated by *Allosyncarpia ternata* closed forest. Potential changes to the aeration regime of the soil is most likely at the boundaries between laterites and sands/clays of the riparian zone. Increased waterlogging is unlikely on the lateritic soils which support open forest, as these soils are generally well-drained.

Increased production of biomass coupled with higher rainfall and temperatures would inevitably lead to greater rates of decomposition of organic matter and rates of nutrient cycling (Sombroek, 1990). Nitrogen availability may therefore increase. Termites, which are major herbivores in Australia's savannas may also be favoured and their role as concentrators of nutrients enhanced.

Major changes to the water regimes of the rivers and associated floodplains are likely if sea level rises, which would have substantial consequences for the associated biota. Rivers have a tidal range of 3-5 m even 50 km inland, and salt water intrusion onto the floodplains would cause the death of species adapted to fresh water such as sedges, native rice (*Oryza* spp.) and would cause an expansion of the mangrove communities. In particular, rapid expansion of tidal creeks poses the major threat to the freshwater wetlands (Knighton *et al.*, 1991).

Figure 9.
Idealised profile along topographic-soil gradient between ridge sites and seepage sites at Kapalga, Kakadu National Park. Elevation = height above sea level. Upper line shows position of land surface; lower line indicates depth to decomposing parent material. Where this coincides with a change in slope, there is a zone of active erosion.

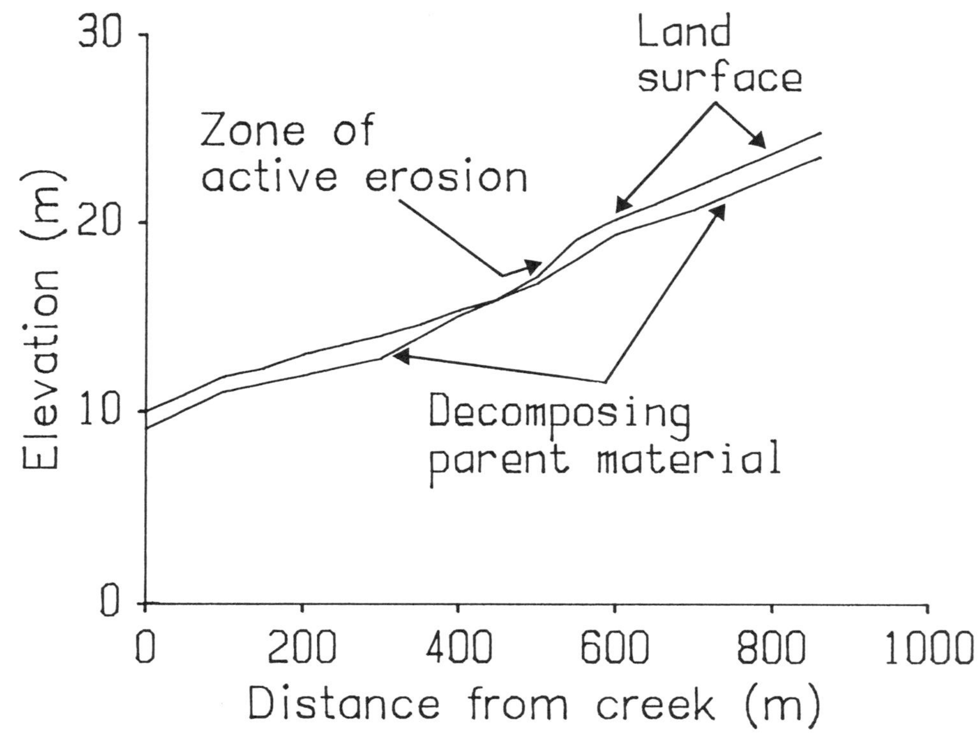

The composition of the floodplain vegetation is affected strongly by depth and persistence of floodwaters (Table 5) and an increase in the frequency/persistence of fresh water may cause a contraction in those species associated with shallower water. Increased rainfall may mean that the vegetation surrounding emphemeral streams and seeps takes longer to cure, and is therefore burnt less frequently, or may be burnt later in the dry season. This may cause a decrease in the abundance of annual compared with perennial grasses.

Table 5.
Relationship between vegetation dominants and water depth on floodplains at Kapalga, Kakadu National Park. Source: Dr J.A. Taylor, CSIRO Division of Crops and Pastures, Brisbane.

Mean water depth (cm)	Dominant species
0	*Fimbristylis littoralis*
10	*Oryza* sp./*Ipomoea aquatica*
40	*Eleocharis sundaica*/*E. dulcis*/*Oryza* sp.
70	*Eleocharis sphacelata*
100	Open water/*Nymphaea violacea*

The degree of deciduousness of the woody species of the savannas during the dry season may decrease, although we have no quantitative studies on rates of leaf turnover. Semi-deciduous species, such as the major eucalypts, are those most likely to show such a response. The response of fully deciduous trees such as *Terminalia* is difficult to predict, because we do not know the critical levels of soil and/or atmospheric moisture which govern the timing and extent of leaf fall. However, whilst leaf retention may increase early in the dry season, it is highly unlikely that such species will retain their leaves throughout the dry season, as total leaf fall occurs by mid-dry season, even after substantial wet seasons (R.J. Williams, unpublished data). In terms of tree-grass interactions, a leafier canopy during the dry season does not necessarily mean less grass in the understorey, as the grasses do not grow during the dry season. Herbivory in the woody species due to insects is likely to increase with increasing rainfall, given the general positive relationship between the degree of rainfall and herbivory, especially in the following dry season (Figure 5). Much more quantitative data are needed, however, to develop models of herbivory in response to climatic variation.

Any changes in the composition of the vegetation of the wetlands may have detrimental impacts on animal species which have very specific water-depth related resource requirements which they satisfy by high mobility. For example the magpie goose is a major migratory bird of Australia's wet-dry tropics, with extensive numbers (Morton & Brennan, 1991). Its previous range extended over most of Australia, but there has been a continuous decline in numbers since 1900, and currently it is restricted to northern Australia. The species has very specific requirements for both commencement of breeding and survival of young. Nesting occurs predominantly in moderately dense *Eleocharis* and *Oryza* grasslands of the floodplains, up to 1.2 m tall, which are flooded annually to depths of 0.3-0.9 m. The goslings feed mainly on wild rice seeds (Frith & Davies, 1961). In terms of breeding, mean annual rainfall is less important than the distribution of rain within the wet season. Breeding is most successful when at least 250 mm of rain falls per month for three successive months. Neither the total rainfall nor the onset of the rains appear to be important determinants of breeding success (Corbett, 1988). Thus the increase in total rain associated with climate change will have little effect on breeding numbers. Of far greater concern is the possibility of more frequent flooding of the floodplains due to rising sea levels. The majority of present breeding areas would be inundated, with potentially disastrous consequences. Changes to the composition of floodplain vegetation as consequence of saltwater intrusion, especially a decrease in the amount of *Oryza* (a major dietary component), will be detrimental. The potential for relocating breeding further inland is unknown, but is likely to be minimal.

Reptiles in which gender is determined by the temperature at which eggs are incubated are well-adapted to fluctuating environmental conditions, and would cope behaviourally with the small changes in temperature which are predicted by the GCMs. The salt water, or estuarine crocodile *Crocodylus porosus* is the major predator of the large river systems, and is found in both salt and fresh water. Crocodiles do not posses sex chromosomes, and the temperature at which eggs develop determines the gender. In addition, temperature of incubation affects embryo mortality (Webb *et al.*, 1986; Webb & Cooper-Preston, 1989). The temperature range which determines both survival and gender is narrow; at 31°C hatchlings are mostly female, at 32° C they are all male, at 33°C all female. Severe mortality occurs at 34°C, with survivors mostly female; 35°C is 100% lethal. Eggs are laid at the beginning of the wet season (November-December), and nest numbers are highly correlated with the Southern Oscillation Index in June/July, suggesting that females respond to environmental conditions well in advance of the onset of the breeding season. The nest sites during incubation are subject to variation in temperature, due to variation in ambient temperature, and flooding of nests. However, crocodiles can compensate very precisely for variations in temperature. The current range of variation in ambient temperatures experienced by developing eggs is similar to that predicted under the scenario of change, and crocodiles are therefore likely to compensate behaviourally, as they do currently. For fresh water crocodiles *Crocodylus johnstonii* the scenario is similar,with behaviour compensating for any climate change, although the temperature range over which gender and survival is determined is narrower than for estuarine crocodiles. Crocodiles, it must be said, have been coping with climate change for 200 million years.

Of the larger mammals, the agile wallaby *Macropus agilis* is the most widespread and abundant macropodid in the wet-dry tropics of Australia. Like the red kangaroo *M. rufus*, the dominant macropodid of the arid regions, the agile wallaby is opportunistic in breeding strategy, which is an adaptation to environmental uncertainty. The greatest uncertainty which the agile wallaby faces is not lack of rainfall, but too much rain falling too suddenly. Agiles wallabies breed continuously in the dry season, but the proportion of females which cease to breed in the wet season is inversely related to the amount of dry season feeding grounds which are flooded in the wet (Bolton *et al.*, 1985). Thus flooding renders food unavailable. With increased chance of severity of rainfall events with changing climate, wallaby breeding may decrease, due to increased extent of flooding of feeding grounds.

We stress, however, that at these scales both predicting and attributing causal links between observed vegetation change and climate change is difficult due to substantial variation in the vegetation in response to variable conditions of other factors, particularly fire. For example, single, infrequent intense fire may have long-lasting effects because of the differential effects of such fires on mortality and regeneration within the woody species. The carry-over effects of this onto animals may be substantial.

Introduced species

The proportion of introduced species within the biota of the savannas of northern Australia is relatively low; some 6% of the 1400 plant species in Kakadu National Park are introduced (Cowie & Werner, 1987). Along catchment gradients within the Alligator Rivers region, the marginal communities (some eucalypt woodlands, monsoon forests, and riparian lowlands of the seepage zones) appear to be more sensitive to invasion of alien plants than are either the wetlands, or savanna forests and woodlands (Figure 10; data from Cowie & Werner, 1987). Two alien species which have extensive distributions in northern Australia, and which may have detrimental impacts on the native biota, are the woody weed *Mimosa pigra* and the cane toad *Bufo marinus*. Both species may be sensitive to climatic change.

Mimosa is a weed of the extensive flood plains of the major river systems. Its seeds, which may number 12,000 m^{-2} in the top 10 cm of the soil (Lonsdale *et al.*, 1988) are spread by flood waters. Any increase in flooding will therefore increase the rate of spread of seed, and is likely to increase the rate of expansion of mimosa stands, as expansion is positively related to the total rainfall of the previous wet season (W.M. Lonsdale, pers. comm.). It is unlikely, however, that mimosa will invade the lighter, better-drained soils of the forests and woodlands, however, should annual rainfall increase.

The cane toad, introduced in tropical Queensland in the 1930s, has expanded its range into the Northern Territory recently (Freeland & Martin, 1985). In its area of origin (Central and South America), lack of prey and dehydration during the dry season are significant factors affecting mortality (Zug & Zug, 1979). In the wild, toads can survive potentially lethal temperatures by selection of more favourable micro-habitats. Increased total annual rainfall will probably assist the spread of cane toads, by providing more prevalent moister habitats across the present, relatively arid barrier of the Gulf of Carpentaria region. The predicted increases in temperatures are below the lethal limits to both tadpoles and adults.

Monitoring change

Monitoring the fate of savannas in the long term depends on the maintenance of long term reference sites. We are maintaining several such areas. At Kapalga, within Kakadu National Park, we are conducting a long-term, landscape-scale experiment on the role of fire and water in tropical savannas. We have also established a continental-scale transect, with permanent reference sites, between Darwin and Tennant Creek (around 1,000 km). Each of these projects provide opportunities to further define the major determinants of tropical savannas (moisture, soil nutrients, fire and herbivory), monitor long-term changes and test hypotheses concerning biotic responses to climate change. In conjunction with monitoring, there is a need to perform manipulative experiments testing these key ideas. In particular, the wet-dry tropics are ideal for testing the differential responses between C_3 and C_4 plants to elevated CO_2 at the scale of the whole plant in the field.

Figure 10.
Species richness of introduced plants as a function of habitat (wetland, margin, savanna) for Kakadu National Park. Data from Cowie and Werner (1986). Communities are *Fimbristylus* sedgeland (Fim); herb swamp (HS); *Melaleuca* swamp (Mel); *Pandanus* lawn (Pan); margin whitegum woodland (WgW); monsoon vine forest (MVF); riparian lowland of seepage zones (Rip); eucalypt woodland (EuW), eucalypt forest (EuF).

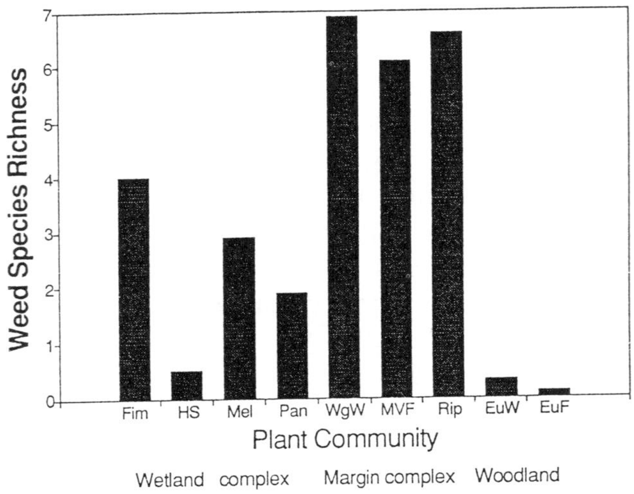

Phenological behaviour is likely to be a sensitive index of response in the vegetation to patterns of inter- and intra-annual rainfall varibility. Major experimental projects addressing the issue of phenology are currently being established at a number of study sites in the humid savannas of the Darwin region with Drs D. Eamus and G. Duff at the Northern Territory University.

Conclusions

The biota of wet-dry savannas has evolved in response to conditions of high stress and frequent disturbance (Walker & Menaut, 1988), and high climatic variability. The predicted changes for the the most important determinant of savanna function, rainfall, are small compared with both the spatial and temporal patterns which occur at present within intact ecosystems. To detect the predicted changes in rainfall will require 50-300 years of observations. Changes in total rainfall are therefore unlikely to affect substantially the structure of savannas of Australia's wet-dry tropics at the sub-continental scale. Of greatest importance to both plants and animals will be changes to the distribution of rainfall within seasons, which at present are not predictable. The impact of elevated CO_2 is difficult to predict, but there is the potential for an increase in the woody C_3 component over the grassy C_4 component of the vegetation. This has the potential to alter fire regimes. Within the landscape there are particular zones and species which would be more sensitive to climatic change. These include the floodplains of the northern rivers which may suffer inundation by saltwater, and the landscape barriers to southward migration of species. In all communities, productivity may increase, but major compositional changes are less likely to be affected by changes in mean conditions, as they appear to depend more on subtle changes in temporal pattern of rain within seasons than mean differences between seasons. The capacity of vegetation to adjust to potential climatic changes will be lessened by current processes which are known to degrade landscape at large scales, in particular the invasion of exotic species, and grazing by domestic livestock.

Acknowledgements

We thank G. Duff, D. Eamus, M. Lomsdale, B. Walker, D. Bowman, R. Fensham, and G. McAdam for constructive criticism of the manuscript. We thank the Australian National Parks and Wildlife Service for logistical support. This is T.E.R.C. publication No. 799.

References

Andersen, A.N. & W.M. Lonsdale. 1990. Herbivory by insects in Australian tropicalsavannas: a review. *J. Biogeog.* **17**: 433-444.

Andrew, M.H. & J.J. Mott. 1983. Annuals with transient seed banks: The population biology of indigenous *Sorghum* species of tropical north-west Australia. *Aust. J. Ecol.* **8**: 265-276.

Archer, M. & B. Fox. 1984. Background to vertebrate zoogeography in Australia. In: Archer M. & G. Clayton (eds) *Vertebrate Zoogeography and Evolution in Australasia*. Hesperian Press, Sydney.

Billings W.D., J.O. Luken, D.A. Mortensen & K.M. Peterson. 1983. Increasing atmospheric carbon dioxide: possible effects on Artic tundra. *Oecologia (Berlin)* **58**: 286-289.

Bolton, B.L., A.E. Newsome & J.C. Merchant. 1985. Breeding in the agile wallaby: opportunistic breeding in a seasonal environment. *Proc. Ecol. Soc. Aust.* **13**:73-79.

Bowman, D.M.J.S. 1986. Stand characteristics, understorey associates and environmental correlates of *Eucalyptus tetrodonta* F. Muell. forests on Gunn Point, northern Australia. *Vegetatio* **65**:105-113

Bowman, D.M.J.S., B.A. Wilson & R.J. Hooper. 1988a. Response of *Eucalyptus* forest and woodland to four fire regimes at Munmarlary, Northern Territory, Australia. *J. Ecol.* **76**: 215-232.

Bowman, D.M.J.S., B.A. Wilson & C.R.Dunlop. 1988b. Preliminary biogeographic analysis of the Northern Territory vascular flora. *Aust. J. Bot.* **36**: 503-517.

Bowman, D.M.J.S. & L. McDonough. 1991. Tree species distribution across a seasonally flooded elevation gradient in the Australian monsoon tropics. *J. Biogeogr.* **18**: (in press).

Bradshaw, A.D. & T. McNielly. 1991. Evolutionary response to global climate change. *Ann. Bot.* **67** (Supplement 1): 5-14.

Braithwaite, R.W. 1987. Effects of fire regimes on lizards in the wet-dry tropics of Australia. *J. Trop. Ecol.* **3**: 265-75.

Braithwaite, R.W. 1990. Australia's unique biota: implications for ecological processes. *J.Biogeog.* **17**: 347-354.

Braithwaite, R.W. 1991. Aboriginal fire regimes of monsoonal Australia in the 19th century. *Search* **22**: 247-249.

Braithwaite, R.W. & J.A. Estbergs. 1985. Fire patterns and woody vegetation trends in the Alligator Rivers region of North Australia. In: J.C. Totthill & J.J. Mott (eds) *Ecology and Management of the World's Savannas*. Australian Academy of Science, Canberra.

Braithwaite, R.W. & J.A. Estbergs. 1987. Firebirds of the Top End. *Aust. Nat. Hist.* **22**: 298-302

Braithwaite, R.W. & Werner P.A. 1987. The biological value of Kakadu National Park. *Search* **18**: 296-301.

Burbidge, N.T. 1960. The phytogeography of the Australian region. *Aust. J. Bot.* **8**: 75-211.

Busby, J.R. 1988. Potential implications of climate change on Australia's flora and fauna. In: G.I. Pearman (ed) *Greenhouse: Planning for Climate Change.* Brill Pub. Coy. Leiden.

Chappell, J. & C.D. Woodroffe. 1985. Morphodynamics of Northern Territory tidal rivers and floodplains. In: K.N. Bardsley, J.D.S. Davie & C.D. Woodroffe (eds) *Coasts and Tidal Wetlands of the Australian Monsoon Region.* Australian National University Northern Australia Research Unit, Mangrove Monograph No. 1, Darwin.

Cogger, H.G. 1988. *Reptiles and Amphibians of Australia.* 5th Edition. Reed, Sydney.

Cook, G.D. & M.H. Andrew. 1991. The nutrient capital of indigenous *Sorghum* species and other understorey components of Savannas in north-western Australia. *Aust. J. Ecol.* **16 (3)**: 375-384.

Corbett, L.K. 1988. *Fluctuations in the yearly numbers of magpie goose (*Anseranas semipalmata*) nests at Kapalga in Kakadu National Park.* Report to Australian National Parks and Wildlife Service, Canberra.

Cowie, I.D. & Werner, P.A. 1987. Weeds in Kakadu National Park - A Survey of Alien Plants. Report to Australian National Parks and Wildlife Service, Canberra.

CSIRO. 1990. *Regional impact of the enhanced greenhouse effect on the Northern Territory.* CSIRO, Melbourne.

Dunlop, C.R. & D.M.J.S. Bowman. 1986. Atlas of the vascular plant genera of the Northern Territory. *Australian Flora and Fauna Series No. 6*, Australian Government Publishing Service, Canberra.

Eamus, D. 1991. The interaction of rising CO_2 and temperature with water use efficiency. *Plant, Cell Environm.* **14**

Eamus, D. & P.G. Jarvis. 1987. The direct effects of increase in the global atmospheric CO_2 concentration on natural and commercial temperatre trees and forests. *Adv. Ecol. Res.* **19**: 1-55.

Fensham, R.J. 1990a. *Regeneration in tropical eucalypt forest on Melville Island.* PhD Thesis, University of Tasmania.

Fensham, R.J. 1990b. Interactive effects of fire frequency and site factors in tropical eucalypt forest. *Aust. J. Ecol.* **15**: 255-266.

Freeland, W.J. & K.C. Martin. 1985. The rate of range expansion by *Bufo marinus* in northern Australia, 1980-1984. *Aust. Wild. Res.* **12**: 555-559.

Frith, H.J. & S.J.J.F. Davies. 1961. Ecology of the magpie goose. *CSIRO Wildlife Research* **6**: 91-141.

Gill A.M., L. Belbin & G.M. Chippendale. 1985. Phytogeography of *Eucalyptus* in Australia. *Australian Flora and Fauna Series, No. 3.*

Gifford, R.M. 1988a. Interactions with vegetation. In: G.I. Pearman (ed) *Greenhouse: Planning for Climatic Change.* CSIRO, Melbourne.

Gifford, R.M. 1988b. Direct effects of higher carbon dioxide concentrations on vegetation. In: G.I. Pearman (ed) *Greenhouse: Planning for Climatic Change.* CSIRO, Melbourne.

Gregory, S. 1991. Spatial patterns of annual July-June rainfall fluctuations in tropical Queensland, 1900/01 to 1979/80. Australian Geographic Studies **29**: 93-113.

Hall, D.O. & Scurlock, J.M.O. 1991. Climate change and productivity of natural grasslands. *Ann. Bot.* **67** (Supplement 1): 49-55.

Hattersley, P.W. 1983. The distribution of C_3 and C_4 grasses in Australia in relation to climate. *Oecologia* (Berlin) **57**: 113-128.

Henderson-Sellers, A. 1991. Policy advice on greenhouse induced climatic change: the scientist's dilemma. *Progr. Phys. Geogr.* **15**: 53-70.
Hnatiuk, R.J. & L. Pedley. 1985. Biogeography of *Acacia* in northern Australia. *Proc. Ecol. Soc.* **13**: 33-42.
Holland, G.J. 1986. Interannual variability of the Australian summer monsoon at Darwin. *Month. Weather Rev.* **114**: 594-604.
Holland, G.J., J.L. McBride & N. Nicholls. 1988. Australian region tropical cyclones and the greenhouse effect. In: Pearman, G.I. (ed) *Greenhouse: Planning for Climate Change.* CSIRO, Melbourne.
Huntley, B. 1991. How plants respond to climate change: Migration rates, individualism and the consequences for plant communities. *Ann. Bot.* **67** (Supplement 1): 15-22.
Houghton, J.T., G.J. Jenkins & J.J. Ephraums. Eds. 1990. *Climate Change: The IPCC Scientific Assessment.* Intergovernmental Panel on Climate Change. Cambridge University Press.
James, C.D. & R. Shine. 1985. The seasonal timing of reproduction: a tropical-temperate comparison in Australian lizards. *Oecologia* (Berlin) **67**: 464-474.
Kerle, J.A. 1985. Habitat preference and diet of the northern brushtail possum *Trichosurus arnhemansis* in the Alligator Rivers region, Northern Territory. *Proc. Ecol. Soc. Aust.* **13**:161-176.
Kennedy, M. 1991. *Australia's Endangered Species.* Ausworld Publishing, Australia.
Knighton, A.D., K. Mills & C.D. Woodroffe. 1991. Tidal-creek extension and saltwater intrusion in northern Australia. *Geology* **19**: 831-834.
Lonsdale, W.M. & R.W. Braithwaite. 1991. Assessing the effects of fire on vegetation in tropical savannas. *Aust. J. Ecol.* **16**:363-374.
Lonsdale, W.M., K.L.S. Harley & J.D. Gillett. 1988. Seed bank dynamics in *Mimosa pigra*, an invasive tropical shrub. *J. Appl. Ecol.* **25**: 963-976.
Love, G. 1988. Cyclone storm surges: post greenhouse. Pp 202-215 in: G.I Pearman (ed) *Greenhouse: Planning for Climatic Change.* CSIRO, Melbourne.
Miles, J. 1979. *Vegetation Dynamics.* Chapman and Hall, London.
Moffatt, I. 1991. Possible ecological impacts of the greenhouse effect on the Northern Territory, Australia. *Glob. Ecol. Biogeogr. Let.* **1**: 102-107.
Mollah, W.S., W. de Launey & M.A. Haynes. 1991. Long-term characteristics of seasonal rainfall at Katherine, Northern Territory. *Austr. Geogr. Stud.* **29**: 71-92.
Moore, P.D. & S.B. Chapman. 1986. *Methods in Plant Ecology.* Blackwell Scientific Publications, Oxford.
Morton, S.R. & K.G. Brennan. 1991. Birds. In: C.D. Haynes, M.G. Ridpath & M.A.J. Williams (eds) *Monsoonal Australia.* Balkema Publishers, Rotterdam.
Mott, J.J., J. Williams, M.H. Andrew & A.N. Gillison. 1985. Australian savanna ecosystems. In: J.C. Totthill & J.J. Mott (eds) *Ecology and Management of the World's Savannas.* Australian Academy of Science, Canberra.
Northcote, K.H. 1968. *Atlas of Australian Soils. Explanatory notes for Sheet 8, Northern Part of Northern Territory.* CSIRO, Melbourne University Press, Melbourne.
Ojima, D.S., T.G.F. Kittel, T. Rosswall & B.H. Walker. 1991. Critical issues for understanding global change effects on terrestrial ecosystems. *Ecol. Applic.* **1**: 316-325.
Pittock, A.B. 1988. Actual and anticipated changes in Australia's climate. In: G.I. Pearman (ed) *Greenhouse: Planning for Climate Change.* CSIRO, Melbourne.

Pittock, A.B. & H.A. Nix. 1986. The effect of changing climate on Australian biomassproduction - a preliminary study. *Clim. Change* **8**: 243-255.

Press, A.J. 1988. Comparisons of the extent of fire in different land management systems in the top end of the Northern Territory. *Proc. Ecol. Soc. Aust.* **15**: 167-175.

Ridpath, M. 1985. Ecology in the wet-dry tropics: how different? *Proc. Ecol. Soc. Aust.* **13**:3-20.

Rosenzweig, C. & R. Dickinson. 1986. Climate-Vegetation Interactions. Office for Interdisciplinary Earth Studies, University Corporation for Atmospheric Research. Boulder, Colorado.

Russell, J.S. 1981. Geographic variation in seasonal rainfall in Australia - an analysis of the 80 year period 1895-1974. *J. Aust. Inst. Ag. Sci* **47**:59-66.

Russell-Smith, J. 1991. Classification, species richness, and environmental relations of monsoon rain forest in northern Australia. *J. Veg. Sci.* **2**: 259-278.

Sarmiento, G. & M. Monasterio. 1983. Life forms and phenology. Pp 79-108 in: Bourliere, F. (ed) *Ecosystems of the World 13: Tropical Savannas.* Elsevier, Amsterdam.

Sombroek , W.G. 1990. Soils on a warmer earth: tropical regions. In: Scharpenseel, H.W., M. Scomaker & A. Ayoub (eds) *Soils on a Warmer Earth.* Elsevier Amsterdam.

Spencer, B. 1896. *Report on the work of the Horn Expedition to Central Australia. Part 2. Zoology.* Melbourne.

Strachan, R. 1983. *The Complete Book of Australian Mammals.* Angus and Robertson, Sydney.

Taylor, J.A. & C.R. Dunlop. 1985. Plant communities of the wet-dry tropics of Australia: the Alligator Rivers region, Northern Territory. *Proc. Ecol. Soc. Aust.* **13**: 83-127.

Taylor, J.A. & D. Tulloch. 1985. Rainfall in the wet-dry tropics: extreme events at Darwin and similarities between years during the period 1870-1983 inclusive. *Aust. J. Ecol.* **10**: 281-295.

Walker, B.H. & J-C Menaut. 1988. Responses of Savannas to Stress and Disturbance.*Publication No. 1. Research Proceedure and Experimental Design for Savanna Ecology and Management.* IUBS/UNESCO.

Webb, G.J.W., A.M. Beal, S.C. Manolis & K.E. Demsy. 1986. The effects of incubation temperatures on sex determinatoin and embryonic development rate in *Crocodylus johnstoni* and *C. porosus*. In: G.J.W. Webb, S.C. Manolis & P.J. Whitehead (eds) *Wildlife Management: Crocodiles and Alligators.* Surrey Beaty and Sons.

Webb, G.J.W. & H. Cooper-Preston. 1989. Effects of incubation temperature on crocodiles and the evolutution of of reptilian oviparity. *Amer. Zool.* **29**: 953-971.

Werner, P.A. 1986. Population dynamics and productivity of selected forest trees in Kakadu National Park. Report to Aust. National Parks and Wildlife Service.

Williams, M.A.J. 1991. Evolution of the landscape. In: C.D. Haynes, M.G. Ridpath, & M.A.J. Williams (eds) *Monsoonal Australia.* Balkema Publishers, Rotterdam.

Wilson, B.A., P.S. Brockelhurst, M.J. Clark & K.J.M. Dickinson. 1990. Vegetation Survey of the Northern Territory, Australia. Explanatory notes to accompany 1: 1,000,000 Map Sheets. Conservation Commission of the Northern Territory, Darwin.

Woinarski, J.C.Z. 1990. Effects of fire on the bird communities of tropical woodlands and open forests in northern Australia. *Aust. J. Ecol.* **15**: 1-22.

Woinarski, J.C.Z. & R.W. Braithwaite. 1990. Conservation foci for Australian birds and mammals. *Search* **21**:65-68.

Woinarski, J.C.Z. & R.W. Braithwaite. 1991. *Wildlife of Kakadu Stage III: A synthesis.* Report to Australian National Parks and Wildlife Service, Canberra.

Woodward, F.I. 1987. *Climate and Plant Distribution.* Cambridge University Press, Cambridge.

Woodward, F.I. 1992. A review of the effects of climate on vegetation: ranges, competition and composition. Pp 105-123 in: R.L. Peters & T.E. Lovejoy (eds) *Global Warming and Biological Diversity*. Yale Univ Press, Newhaven.

Zug, G.R. & P.B. Zug. 1979. The marine toad *Bufo marinus*: a natural history resume of native populations. *Smithsonian Contrib. Zool.* **24**: 1-58.

Possible vegetation shifts in Siberia under climatic change

Nadja M. Tchebakova

International Institute for Applied System Analysis, IIASA, A-2361 Laxenburg, Austria; and Institute of Forest and Wood, Russian Academy of Sciences Academgorodok, 660036, Krasnoyarsk, Russia

Robert A. Monserud

Intermountain Research Station, Forest Service, United States Department of Agriculture 1221 S. Main St., Moscow, Idaho 83843, USA

Rik Leemans

Global Change Department, National Institute for Public Health and Environmental Protection, PO Box 1, 3720 BA Bilthoven, The Netherlands

Dina I. Nazimova

Institute of Forest and Wood, Russian Academy of Sciences Academgorodok, 660036, Krasnoyarsk, Russia

Abstract

A regional vegetation model for Siberia is used to evaluate possible changes in major vegetation types resulting from projected climate change. The vegetation model predicts vegetation type as a function of temperature sums, a dryness index, and a continentality index. Climate change scenarios are predicted by four General Circulation Models (GCMs) of the atmosphere (GFDL, GISS, OSU, and UKMO), assuming a doubling of CO_2. Results indicate the potential for enormous change in the vegetation of Siberia as a result of climate change. All four GCMs predict large reductions in Tundra and Forest-Tundra, reductions and northern shifts in the various classes of Taiga, and reductions in Steppe. Large increases are predicted for the Temperate Forest Steppe and Temperate Steppe vegetation zones, and for Forest Steppe and Subtaiga in the Boreal zone. Generally, vegetation bands are predicted to shift northward, with extensive areas of temperate vegetation introduced in the south. Changes in mountainous areas are more complex owing to the greater variety of environments, but generally vegetation zones will increase in altitude. Kappa statistics for judging agreement between the current climate vegetation map and the climate change predictions indicate that there is essentially no agreement between maps, thus confirming the visual impression of wholesale change in the predicted vegetation of Siberia.

Introduction

The possibility of significant climate change on a global scale is being taken increasingly seriously by the scientific community (eg. Houghton *et al.*, 1990; Adams *et al.*, 1990). A global warming induced by increased atmospheric greenhouse gases has been projected by most of the important Global Circulation Models (GCMs) (see Harrison, 1990). Evaluation of the effects of global warming simulated by the GCMs indicate great shifts in the size and location of vegetation zones (Emanuel *et al.*, 1985; Monserud & Leemans, 1992; Tchebakova *et al.*, 1992).

There have been few regional studies of the effect of possible climate change on the distribution of vegetation (eg. Solomon, 1986; Prentice *et al.*, 1991). Regional vegetation models afford much greater accuracy and resolution than global vegetation models for the same region. However, such analyses are handicapped by the lack of correspondingly accurate regional climate change predictions. Although the GCMs produce reliable global distributions of temperature and precipitation, on regional scales there are significant errors in these variables in all models (Houghton *et al.*, 1990). Climate change prediction is basically a global rather than a regional problem, even if the area of interest is only regional. Generally, improvements in regional climate change scenarios are dependent on improvements in the GCMs. Recent work by Giorgi *et al.* (1990) nesting a regional climate model for Europe in a GCM shows promise for improving regional climate predictions, however. Similarly to Solomon (1986), our strategy is to study regional climate change effects by coupling the GCMs' climate change scenarios with a regional vegetation model.

Siberian vegetation model. Our study of possible Siberian vegetation changes under climate warming is based on the Siberian model of Nazimova *et al.* (1990). This model predicts the major vegetation types of the Siberian subcontinent from three parameters: temperature sums (base 10°C), dryness index (**DI**), and continentality index (CI).

Temperature sums characterise the heat resources of climates in terms of plant requirements for growth. As accepted in the Russian literature, temperature sums ($\mathbf{TS_b}$) are calculated as the sum of all positive temperatures (**T**) occurring for the period with daily temperature above a given threshold (**Tb**):

$$TS_b = \int T\, dt \qquad (1)$$

where integration is over the time period with **T > Tb**. In contrast to this is the closely related concept of Growing Degree Days ($\mathbf{GGD_b}$) in Western literature, where only temperatures above the temperature threshold **Tb** considered to be sufficient for plant growth are summed:

$$GDD_b = \int (T - T_b) dt \qquad (2)$$

Therefore, growing degree days are always less than the temperature sums used in Russian literature but strongly correlated with them. For the convenience of Western readers we have converted temperature sums, base 10°, ($\mathbf{TS_{10}}$), used in the Siberian model into $\mathbf{GGD_5}$. The threshold 5° is the minimum temperature required for plant growth in cold climates (Prentice *et al.*, 1992) and usually used in boreal vegetation-climate research. A simple linear regression for the Siberian region explains almost all the variation ($R^2 = 0.993$) between them:

$$GDD_5 = 0.745\ TS_{10} + 12.83 \quad (3)$$

The *Dryness Index* (**DI**) of Budyko (1974, 1986) characterises the dryness (or wetness) of climates and is the ratio of the amount of water that can be evaporated under available heat (potential evaporation) and precipitation. Potential evaporation is estimated from radiation balance, which is calculated from temperature, humidity, and cloudiness (Budyko, 1974). Dryness index is similar to those climatic moisture indices that are based on a ratio of potential evaporation and precipitation; it differs from them primarily in the approach used to calculate potential evaporation. Dryness index strongly correlates to those indices that are based on a ratio of actual and potential evaporation (Zubenok, 1976).

Continentality index (CI) tries to characterise the severity of climates as a function of annual temperature amplitude (July maximum minus January minimum in the northern hemisphere) and latitude (Conrad, 1947). Nazimova *et al.* (1990) showed that Conrad's (1947) CI is of the greatest importance for classifying the vegetation of Siberia with its severe climates. Nazimova *et al.* (1990) classified vegetation not only into thermic zones (eg. polar, boreal, temperate) but into continentality sectors as well (eg. Eastern, Central, and Western Taiga). Sectors are characterised by different vegetation structure within the same thermic zone or subzone.

Siberian vegetation classification

The vegetation classification for Siberia is more detailed than other global systems for this region and consists of 17 classes (Nazimova *et al.*, 1990). Taiga, the main boreal vegetation class, was subdivided into thermic subzones such as Northern, Middle, and Southern Taiga, and then into continentality sectors such as Western, Central, and Eastern. These divisions are well accepted in the Russian literature (Shumilova, 1962; Krylov, 1984; Parmuzin, 1986).

The Western continental, humid sector is characterised by CI values of 40-60. Dark-needled conifer formations (*Pinus sibirica, Abies sibirica, Picea obovata*) dominate in major zonal sites, pine forests (*Pinus sylvestris*) occur in intrazonal sites, and birch (*Betula pendula*) in demutation successions. Within this sector, the Northern Taiga and open woodlands are dominated by *Larix sibirica* and *Picea obovata* of V-Va site quality class; the Middle Taiga is dominated by *Pinus sibirica* and *Picea obovata* of IV (sometimes III) site quality class; and the Southern Taiga contains much *Abies sibirica* of I-III site quality class. Russian site quality literature (eg. Orlov, 1929) uses essentially the same system as in the corresponding American literature (eg. McArdle & Meyer, 1930). Productivity is indexed by site index (dominant height at index age 100 years) and divided into five site quality classes - I is best, III - intermediate and V - worst. Finer classifications may be used: Va and Vb are increasingly poor quality; Ia and Ib increasingly better.

The Central continental, semihumid sector is characterised by CI values of 60-85 (and 60-80 in Steppes and Forest-Steppes). Light-needled formations (*Larix sibirica* and *Pinus sylvestris*) dominate, and dark-needled species are absent in zonal sites. *Picea obovata* and *Pinus sylvestris* occur in intrazonal sites. Within the sector, the Northern Taiga consists of *Larix sibirica* of V-Va site quality class and birch yerniks (dwarf birch thickets), and without pine. The Middle Taiga is dominated by *Larix sibirica* of IV site quality class. The Southern Taiga contains much *Pinus sylvestris* of III site quality class. Dark-needled species are reduced.

The Eastern extreme continental, cryo-semihumid (semiarid) sector is characterised by CI values more than 80-85. *Larix* (three species - *L. sibirica, L. gmelinii*, and *L. cajanderii*) absolutely dominates in the absence of *Pinus sibirica* and *Abies sibirica*. Demutation successions mostly occur without birch stages. In montane Taiga there is *Larix sibirica* of III-IV site quality class. Within the Eastern sector, the Northern Taiga contains *Larix cajanderii* of V-Va site quality class; these are probably the coldest forests in the world in the winter. The Middle Taiga is dominated by *Larix gmelinii* of IV-V site quality class. No Southern Taiga subzone exists in the Eastern sector owing to the severity of continental climate.

The same Western, Central, and Eastern continentality sectors were used to subdivide Subtaiga (mixed pine and small-leaved birch and poplar forests), Steppe (grasslands), and the Forest-Tundra and Forest-Steppe ecotones. Tundra was not subdivided in our model because of the shortage of climatic information for these extreme northern regions.

Modification of the model

We simplified the Siberian model of Nazimova *et al.* (1990) by defining linear borders between classes and specifying rectangular classes (Figure 1). Borders specified by Nazimova *et al.* (1990) are irregular and with complex contours. In fact, their borders can not be depicted precisely because there are few or no weather stations at the borders and, moreover, weather stations are often located in such sites as valleys, which are not very characteristic for zonal vegetation.

For this version of the Siberian model we provide one more additional class: Tundra-Steppe, an ecotone. Yurtsev (1986) documented small fragments of Steppe communities in Tundra in big river valleys of the northeastern part of Eurasia. He found features of these Steppes to be relic and widespread in previous glacial epochs. Research on the relationship between climate and vegetation in the Sayan mountains in Southern Siberia (Polikarpov *et al.*, 1986) showed that such contacts between Tundra and Steppe on montane slopes in highlands of arid regions are more common than on plains. Prentice *et al.* (1992) found the same regularities while designing their BIOME model and referred to the transition zone between Tundra and Steppe occurring on a microscale in the montane interior of Alaska. Although we know that this Tundra-Steppe ecotone exists, we need more field data before it can be located accurately in the $\boldsymbol{DI\text{-}GGD_5}$ space illustrated in figure 1.

Figure 1.
The Siberia model, based on an ordination between Budyko's Dryness Index (DI) and growing degree days above a base of 5°C (GGD5).

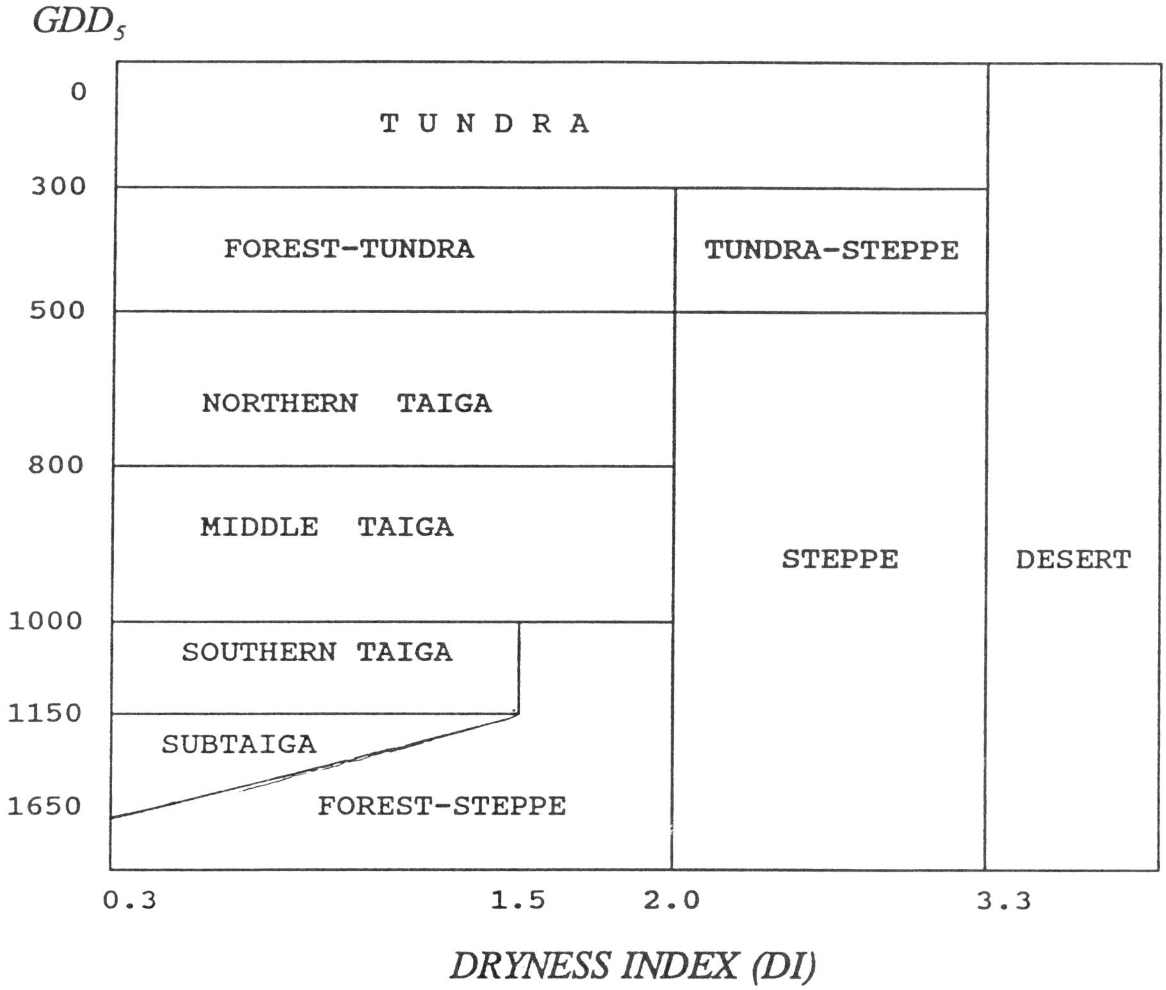

We have added to the Siberian model three more vegetation classes of the temperate zone: Forest-Steppe, Steppe, and Desert. These classes contact with boreal Steppes in the south. These temperate types play an important role in shifting boreal vegetation poleward under warming.

We have not specified the class Ice because it cannot be specified by the available three model parameters used by Nazimova *et al.* (1990). However, it can be modelled quite simply by incorporating a fourth parameter in the model: either July temperature max<0° (Tchebakova *et al.*, 1992) or ***GGD***o=0 (Prentice *et al.*, 1992).

The final modification is that the rectangular transition zone combining Subtaiga and Forest-Steppe classes is divided diagonally (see figure 1) into a less warm and wetter triangular section containing only Subtaiga and a warmer and drier section containing only Forest-Steppe. In West and Central Siberia, Subtaiga represents a narrow transition band between boreal dark-needled Taiga and Forest-Steppe and often cannot be distinguished from Forest-Steppe. In the extreme continental climates of East Siberia there are no such classes as Southern Taiga and Subtaiga. In such cases the transition is from Middle or Northern Taiga directly to Forest-Steppe or Steppe.

Climatic Data

We used current climatic data on temperature, precipitation, and cloudiness from the updated and improved version of the IIASA Climate Database (Leemans & Cramer, 1991); this updated database was generously provided by Wolfgang Cramer (pers. comm.). Humidity (or vapour pressure) was taken primarily from Anonymous (1966-1970). Data from about 1,200 weather stations in Siberia were used to interpolate precipitation on a 0.5° grid, and from about 800-900 stations to interpolate temperature, humidity, and cloudiness. Albedo (necessary to calculate Budyko's dryness index) was taken from Matthews' (1984, 1985) digitised albedo maps for the continents.

To evaluate vegetation shifts in Siberia, we used climatic scenarios of doubling of CO_2, based on four General Circulation Models of the atmosphere (GCMs): **GFDL**, Geophysical Fluid Dynamics Laboratory of NOAA, Princeton (Wetherald & Manabe, 1986; Manabe & Wetherald, 1987); **GISS**, Goddard Institute for Space Studies, NASA, Columbia University (Hansen *et al.*, 1983); **OSU**, Oregon State University, Corvallis (Schlesinger & Zhao, 1989); and **UKMO**, United Kingdom Meteorological Office (Mitchell, 1983; Wilson & Mitchell, 1987). Output of the GCMs is expressed in terms of temperature and precipitation changes (Harrison, 1990). Therefore, we recalculated model climatic parameters based only on changes of these two (see Monserud & Leemans, 1992, for details). The remaining three were taken unchanged from the current climatic sets.

Map comparison

Maps of Siberian vegetation corresponding to current climate and climate change were compared using objective techniques described by Monserud (1990) and Monserud and Leemans (1992). The main comparison tools were the kappa statistic (Cohen, 1960), which measures pixel by pixel agreement between maps, and changes in area. Comparison was also made at a somewhat larger scale (2.5° cells) using a generalised kappa statistic developed by Prentice *et al.* (1992). In addition to changes in area, the distance and direction of shifts in the centroid of the vegetation classes were also calculated (Monserud, 1990).

We used the following "window" to delineate Siberia: the area between 60°E (Ural Mountains) in the west, 140°E (Tchersky Mountains) in the east, 50°N (Mongolian border) in the south, and 75°N (Arctic Ocean) in the north. We also excluded the southeast corner of this window (the vicinity of the Sea of Okhotsk): south of 60°N and east of 140°E. This Siberian window contains 6,891 pixels of 0.5° latitude by 0.5° longitude.

Results

Comparison of the current climate vegetation map with the four GCM climate change scenarios reveals enormous changes in predicted vegetation (cf. figures 2a & 2b). The predicted climate change maps fell into two categories: the GFDL and UKMO scenarios provide similar results (agreement between them is good, with a kappa statistic of 0.60 when using 2.5° cells for comparison) and predict great changes in vegetation. Based on kappa statistics, comparison between the OSU and GISS projection maps also indicate good agreement and somewhat less dramatic changes in vegetation. All four GCMs predict large reductions in Tundra and Forest-Tundra, reductions and northern shifts in the various classes of Taiga, and reductions in Steppe (Table 1). Large increases are predicted for the Temperate Forest Steppe and Temperate Steppe vegetation zones, and for Forest Steppe and Subtaiga in the Boreal zone. Generally, vegetation bands are predicted to shift northward, with extensive areas of temperate vegetation introduced in the south (Table 2). Changes in mountainous areas are more complex, but generally vegetation zones will increase in altitude.

Kappa statistics indicate that there is no agreement between the current climate vegetation map and the four climate change scenarios, thus confirming the visual impression (Figures 2a & 2b) of wholesale change in the predicted vegetation of Siberia. The only vegetation classes predicted by any of the GCMs to have any stable area (ie. remain in the same vegetation class) are Temperate Steppe and small areas of Temperate Forest Steppe and Tundra (Table 1).

GFDL and UKMO Scenarios. According to the GFDL and UKMO climate change scenarios, enormous changes will occur in the boreal zone. Only 5% of Tundra and 15% of Forest-Tundra will survive (percentage change is based on area, not number of cells). Their current areas will be occupied by Northern or even Middle Taiga. Only small areas of Forest-Tundra will be preserved, mainly in cold flat interfluves of continental Siberia such as the Putoran Mountains and Aldan Ridges. Northern Taiga will shift north, and either Middle Taiga or Southern Taiga will occupy its present locations. In Western Siberia, Subtaiga will reach nearly 70°N, and Forest-Steppe will reach to the polar circle, where presently the contact of Northern Taiga and Forest-Tundra is observed. Temperate Forest-Steppe will greatly expand as a result of global warming, reaching as far as 65°N (Figure 2b).

From a small nucleus of Forest-Steppes and Steppes occurring presently among Taiga in Yakutia, GFDL predicts a six-fold increase with climate change. Despite large warming, continentality in Yakutia will not decrease, although in general for Siberia continentality will decrease by 10 units of the Conrad coefficient. For instance, continental eastern variants of Northern and Middle Taiga will disappear. From Forest-Tundra there will remain only about 50% of its western variant and 5% of its central currently dominating variant (Table 1); these areas will survive mainly in the mountains.

Vegetation Zone	Area Comparison (units = 1000 km²) Current Climate Map	Climate Change Maps GFDL	GISS	OSU	UKMO	Percentage Stable Area GFDL	GISS	OSU	UKMO
Forest-Tundra (western)	33.4	15.9	40.2	33.8	18.1	0	0	0	0
Forest Tundra (central)	91.4	4.7	13.4	18.0	.9	0	2	0	0
Forest Tundra (eastern)	2.2	.0	.0	.0	.0	0	0	0	0
Northern Taiga (western)	67.4	38.5	77.9	46.5	48.9	0	1	2	0
Northern Taiga (central)	199.2	34.9	91.8	83.8	9.1	0	6	0	0
Northern Taiga (eastern)	21.6	.0	.0	.9	.0	0	0	0	0
Middle Taiga (western)	54.5	26.8	51.8	36.7	40.8	0	0	0	0
Middle Taiga (central)	117.6	39.4	84.5	103.2	18.8	0	0	0	0
Middle Taiga (eastern)	25.3	.0	.2	3.0	.0	0	0	0	0
Southern Taiga (western)	52.9	17.3	39.2	26.7	21.4	0	0	0	0
Southern Taiga (central)	44.2	45.5	73.5	94.5	34.8	0	0	0	0
Subtaiga (western)	53.8	43.4	102.1	62.6	74.4	0	0	0	0
Subtaiga (central)	22.5	82.9	145.2	159.8	115.5	0	0	0	0
Forest Steppe (western)	50.6	31.4	42.0	32.1	16.3	0	0	3	0
Forest Steppe (central)	15.0	83.4	37.2	57.3	57.1	0	8	1	0
Forest Steppe (eastern)	5.0	30.8	11.9	38.3	1.5	0	20	41	0
Steppe (western)	28.4	2.3	.0	.0	.2	0	0	0	0
Steppe (central)	7.3	3.2	1.9	3.5	.0	3	3	8	0
Steppe (eastern)	5.7	5.5	.6	1.6	.4	50	11	21	0
Temperate Forest Steppe	.2	408.0	149.7	127.5	477.8	100	100	100	100
Temperate Steppe	23.5	132.5	77.2	84.0	102.4	93	86	92	96
Tundra-Steppe	.0	.7	.1	.0	.0	0	0	0	0
Continental Steppe	.0	.0	.0	.0	.0	0	0	0	0
Tundra	119.6	6.0	12.6	30.2	11.1	5	11	25	9
Desert	12.4	.6	.8	9.8	4.1	2	6	74	28
Totals:	1053.8	1053.8	1053.8	1053.8	1053.8	3	5	6	3

Table 1.
Aeea comparisons between the current climate map and the four climate change scenarios, using the Siberia model.
The percentage of the original vegetation zones remaining unchanged (stable area) is also given.

Figure 2a.
The Siberia model map predictions for current climate.

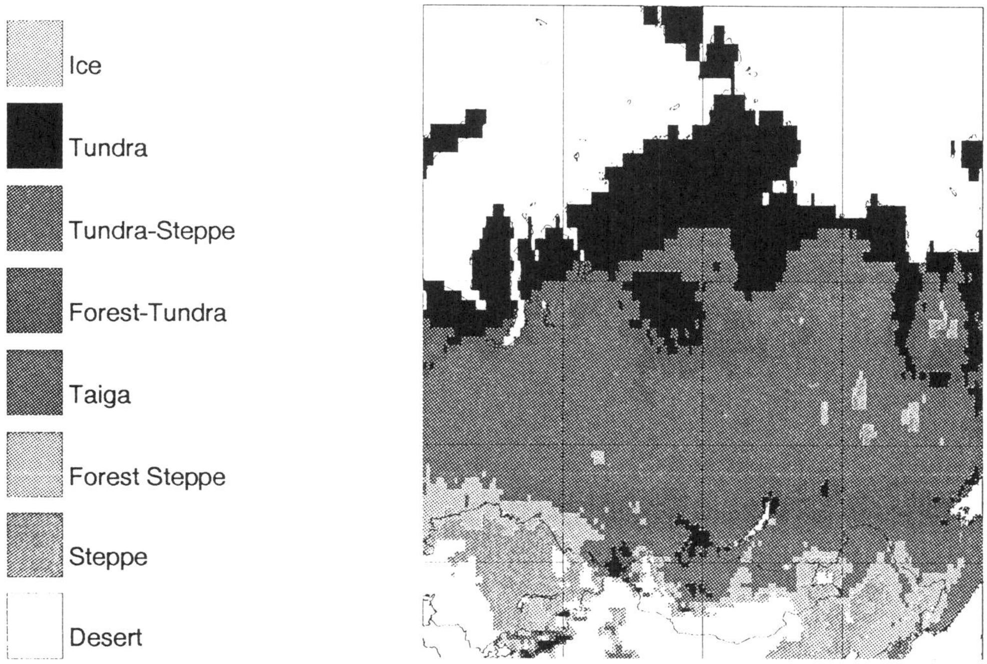

Figure 2b.
The Siberia model map predictions for climate change. This Siberian map is derived from a double CO_2 climate scenario as predicted by the GFDL general circulation model.

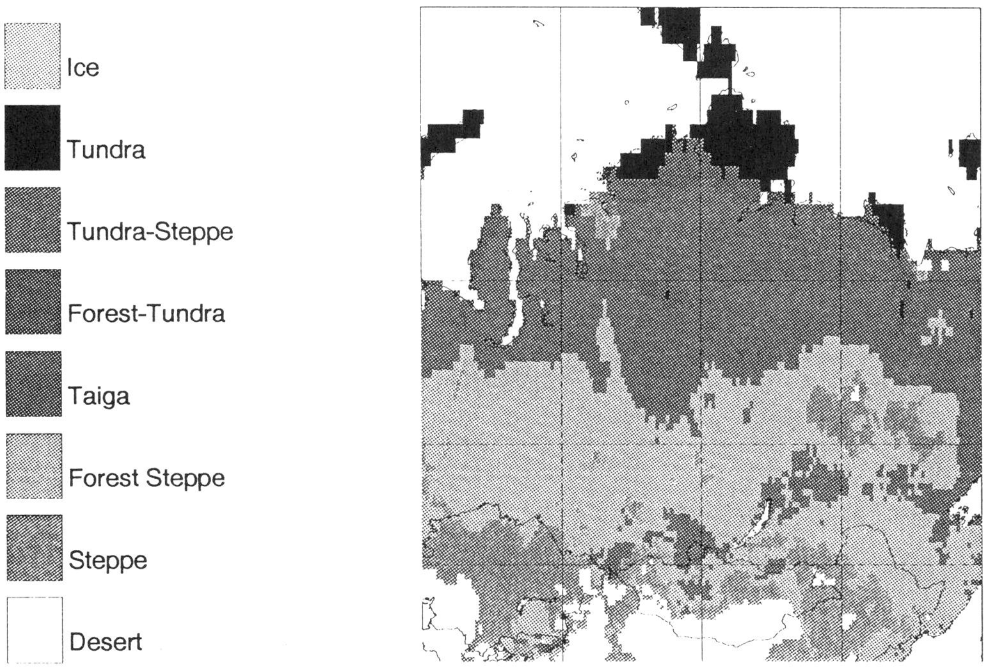

Table 2.
The distance and direction that the median (50th percentile) of each vegetation zone has shifted due to climate change.

	DISTANCE MEDIAN SHIFTED (units = 100 km)				DIRECTION OF SHIFT from base map			
Vegetation Zone	GFDL	GISS	OSU	UKMO	GFDL	GISS	OSU	UKMO
Forest Tundra (central)	3.5	1.2	3.5	3.9	W	NE	NW	W
Forest Tundra (eastern)	.0	.0	.0	.0				
Northern Taiga (western)	8.8	7.5	5.1	11.2	NE	NE	NE	NE
Northern Taiga (central)	9.8	7.3	8.4	9.0	NW	N	NW	NW
Northern Taiga (eastern)	.0	.0	4.6	.0		SE		
Middle Taiga (western)	9.5	6.7	5.7	13.3	NE	NE	N	NE
Middle Taiga (central)	13.8	5.6	8.0	12.4	NW	N	NW	NW
Middle Taiga (eastern)	.0	8.4	6.4	.0		SE	SE	
Southern Taiga (western)	10.0	6.7	6.6	10.7	N	N	NW	N
Southern Taiga (central)	15.1	6.9	9.0	15.7	N	NE	N	N
Subtaiga (western)	11.1	5.0	5.8	9.5	N	N	NW	N
Subtaiga (central)	15.3	9.3	12.1	15.4	NW	NW	NW	NW
Forest Steppe (western)	12.7	9.7	6.5	8.9	NE	E	E	N
Forest Steppe (central)	13.7	2.7	5.4	11.6	NW	NE	NW	NW
Forest Steppe (eastern)	8.0	8.9	11.6	7.0	N	N	NW	NW
Steppe (western)	23.9	.0	.0	24.4	NE			E
Steppe (central)	8.6	8.8	22.5	.0	E	E	E	
Steppe (eastern)	.8	1.7	.4	24.3	SE	NW	NW	W
Temperate Forest Steppe	26.6	32.7	30.4	21.1	W	W	W	W
Temperate Steppe	4.3	2.8	2.8	.9	E	W	W	NW
Tundra-Steppe	99.1	114.0	.0	.0	N	N		
Continental Steppe	.0	.0	.0	.0				
Tundra	6.2	2.8	3.0	3.5	E	N	W	NW
Desert	36.8	.3	.7	1.1	E	NW	W	W

Figure 3 illustrates how the vegetation classes have increased and decreased in area, while statistics presented in Table 2 indicate the main directions and distances of vegetation shifts. Table 2 lists the median (50th percentile), which could be viewed as the centroid of a vegetation class. Except in montane areas, Siberian vegetation classes are fairly homogeneous geographically, enough so that these statistics should project a sufficiently reliable picture (especially because there are no oceans or large bodies of water that would distort the statistics). Under global warming nearly all classes will shift predominantly to the north, although sometimes with an eastern or western component. Only Steppe and Tundra will shift eastward. Tundra and Northern Taiga classes will move 600-1,000 km. Classes of the middle and southern subzones of Taiga will move 1,000-1,500 km north. Boreal Steppe and Temperate Forest-Steppe will invade north as far as 2,500 km.

In warm mountain regions such as the Altai-Sayan mountains, with their complex topography and variety of climates, the full spectrum of vegetation classes can presently be found, including all three continentality variants. The mountains are sufficiently high that one can find large areas of cold altitudinal belts such as Tundra and Forest-Tundra. The UKMO climate projection shows that Tundra and Forest-Tundra will disappear with climate change in such areas. It is known that tree-line in mountains is determined not only by warmth but by wind and lithology as well. Tundra will likely remain in open and rocky highlands unfavorable for tree growth. From montane Taiga forests there will remain only warm variants analogous to flatland Southern Taiga and Subtaiga and small areas of Middle Taiga. Because of large warming in the mountains of Southern Siberia, dark-needled species such as *Abies sibirica* and *Pinus sibirica* might not survive except in refuges in high mountains and northern slopes. The conifers *Pinus sylvestris* and *Larix sibirica* and the deciduous species *Betula pendula* and *Populus tremula* will likely dominate in all vegetation belts, which will increase in elevation up mountains. In some favorable sites the main dark-needled species might be restored through a long-term succession, taking, for instance, 120-140 years for *Pinus sibirica* (Polikarpov *et al.*, 1986).

GISS and OSU scenarios. Vegetation changes predicted by the GISS and OSU scenarios have much in common with the GFDL and UKMO scenarios. Tundra will decrease by 75-90% and survive only near sea shores and islands in the Arctic. Small spots of Tundra will also be found in cold, flat interfluves. Tundra will be mostly replaced by Forest-Tundra. The area of Forest-Tundra on the plain will also decrease, especially its continental variants. Because of decreased continentality the western variant of Forest-Tundra will dominate. In cold mountains Forest-Tundra will generally replace Tundra. Northern and Middle Taiga will survive by 65-70% and will shift far north. Its southern border in the Western sector will move from 60° to 65° N and in the Central and Eastern sectors it will move from 50 to 62°N. The cold variants of Taiga will be replaced by warm variants such as Southern Taiga and Subtaiga. The area of Subtaiga will increase three to four times, but the Southern Taiga will have essentially the same area. In the Eastern sector, Middle Taiga will be replaced by a continental variant of Forest-Steppe, which confirms that in this region continentality of climates will not change much. Under the GISS and OSU projections, Temperate Forest-Steppe will greatly expand, although this is only one-third of the expansion predicted by GFDL and UKMO. Temperate Steppe is projected to triple in size.

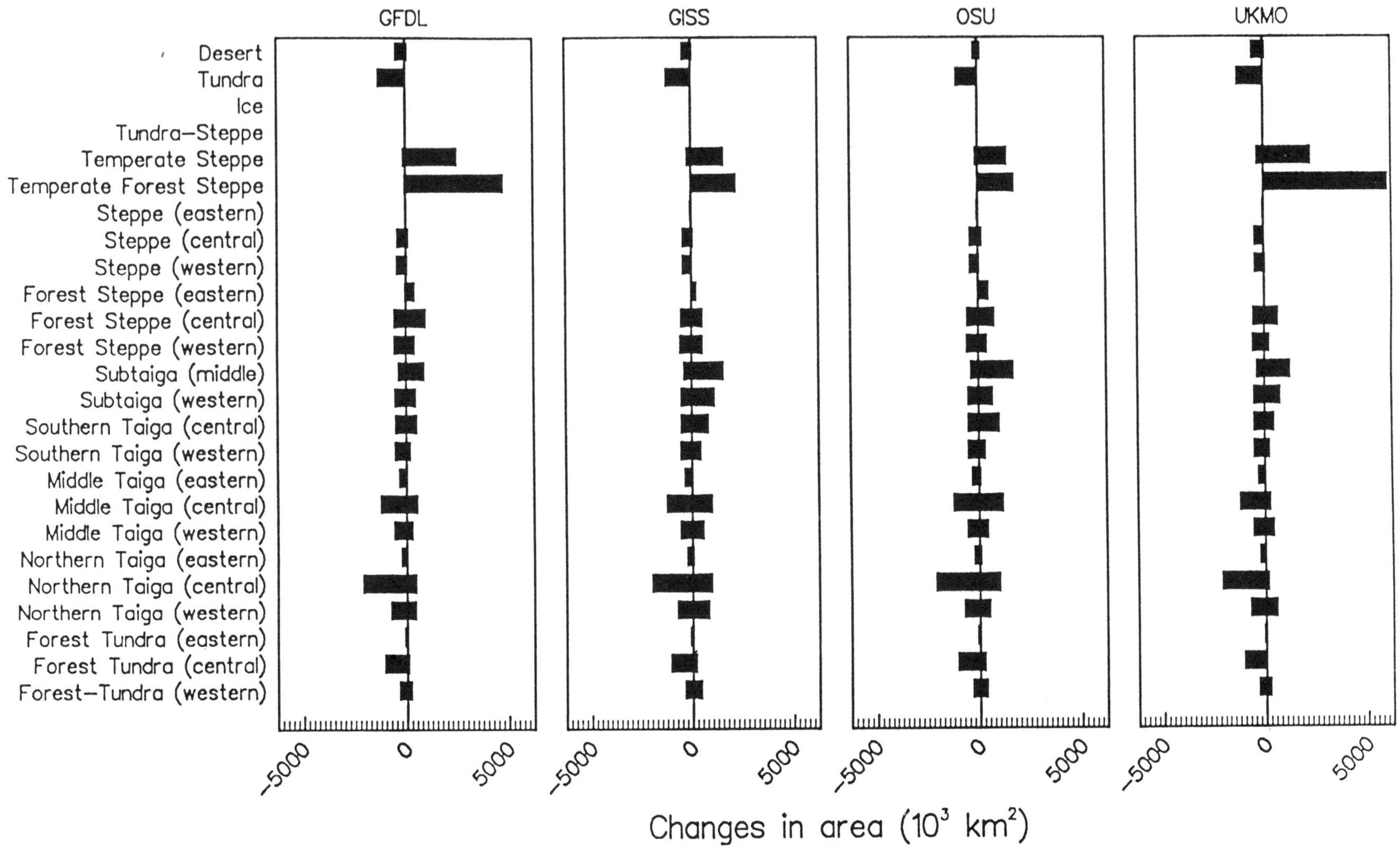

Figure 3.
Change in area by vegetation zones in the Siberia climate change maps for four GCM predictions.
Area to the left of the zero centerline is a reduction (decrease) in area from the current climate vegetation map (Figure 2a).
Area to the right of the centerline is an increase in area into new locations in the climate change map (Figure 2b).

Net change is the increase into new locations (right of centerline) minus the loss of old locations (left of centerline). Under the somewhat more moderate changes predicted by GISS and OSU, distances that vegetation classes shift will be about 1.5 times less than with the GFDL and UKMO scenarios, although the direction of shift (northeasterly) is about the same. Temperate Forest-Steppe is predicted to penetrate into the current boreal zone as much as 3,000 km (Table 2).

The variety of vegetation in relatively warm mountains in Southern Siberia will largely remain, except for such vegetation types as Tundra and Forest-Tundra, which will nearly disappear. Because of climate warming, all altitudinal belts containing forest vegetation will rise upward. Taiga belts will occur in many upland locations currently vegetated by Tundra and Forest-Tundra, and Subtaiga and Forest-Steppe will replace Taiga in lower and middle belts. These general patterns resulting from climate change appear to be more complex in mountains, where there are a wide variety of environments available for vegetation to survive.

In the current climate vegetation map there are five grid cells of the Tundra-Steppe vegetation class, with a total area of 8,000 km^2. This Tundra-Steppe vegetation occurs under specific combinations of cold and dry climates in Yakutia. Under moderate climate warming, Tundra-Steppe vegetation will likely survive but migrate to other places previously occupied by Tundra.

Discussion

Mechanisms of competitive interrelations among species leading to forest succession could be represented on the basis of empirical knowledge accumulated in forestry (Kolesnikov, 1965; Melekhov, 1980; Polikarpov *et al.*, 1986). These interrelations are controlled both by genetic and environmental factors, with climatic factors being among the most important determinants. In spite of a great variety of geomorphology, lithology, and soils, we foresee the following trends in climatic succession in Siberia as a result of global warming.

Under conditions of warming and decreased continentality, forests could become more dense and the forested area could be increased in the Northern Taiga zone. However, in permafrost regions, there is a risk that Northern Taiga will be replaced following warming by yerniks (*Betula nana*, *B. humilis*, *B. middendorfii*) or even by Steppe xerophytic groups if the permafrost melts. The reason is that permafrost provides additional water (in excess of precipitation) from the air by means of condensation and also by melting during a warm season. As an example, one can currently find Taiga in extreme continental climates in Yakutia that otherwise would be too dry to support forests. Global vegetation models based on climatic parameters (Leemans, 1989; Prentice *et al.*, 1991) judge these Northern Taiga permafrost areas as too dry for forest vegetation and instead predict Steppe or even Semideserts.

In the Middle and Southern Taiga of the humid regions, dark-needled species (primarily *Pinus sibirica* and *Picea obovata*, and then *Abies sibirica*) will become more competitive. *Abies* is limited in its ability to penetrate into the Middle Taiga because of permafrost and insufficient soil moisture. Under climatic warming, the course of succession leading to conifer forests will likely occur through a small-leaved (birch) stage, especially in the ecotone between dark-needled humid and light-needled semihumid biomes. Pyrogenic successions leading to domination of birch in forest composition are likely to be widespread. At the next stage under birch canopies, dark-needled undergrowth occurs. Such successions are common on elevated moister areas in Central Siberia and even more so in West Siberia. In the Middle and Southern Taiga zones of the semihumid regions, *Pinus sylvestris* and *Betula pendula* are likely to be active as the most stable and competitive forest-forming species.

In the Southern Taiga and Subtaiga, the pioneer species *Betula pendula* and *Populus tremula* play an important role in disturbed forests. A combination of climate changes between warming and even small increases in precipitation will create especially favorable conditions for *Betula pendula*. On flat interfluves in the Southern Taiga as well as in the Middle Taiga, the potential productivity of birch forests will increase by at least one site quality class (to I-II in the Southern Taiga and to II-III in the Middle Taiga). Under climatic warming in Central Siberia, the ecotone between Middle Taiga and birch-pine Subtaiga will be more common than now.

Because most of the Southern Taiga forests currently in Central Siberia will likely find themselves in a climatic space more suitable for Subtaiga and even boreal Forest-Steppe, natural decline of dark-needled species will inevitably occur on flat interfluves and southern slopes, but they could be preserved in valleys and northern slopes.

To predict dynamic forest succession under global warming we should consider not only climate rates of change but biota response rates to them as well. We should take into account such properties of ecosystems as conservation of biota and soil, competitive abilities of species, and feedback relations within the entire "climate-vegetation-soil" system. In doing this, such accumulated knowledge as multiple sequences of succession as well as ecological and biological characteristics of forest-forming species might be helpful in understanding ecological mechanisms responsible for species migration and vegetation shifts under environmental changes. Ultimately we would need to know the genetic makeup of species to judge their ability to adapt to a changing environment.

Conclusions

Based on a Siberian vegetation model and climate change scenarios (CO_2 doubling) from four GCMs, the vegetation of Siberia is predicted to be almost completely changed as a result of climate warming. Generally, vegetation bands are predicted to shift northward, with extensive areas of temperate vegetation introduced in the south. Changes in mountainous areas are more complex due to a greater variety of environments, but generally vegetation zones will increase in altitude. Large reductions are predicted for Tundra and Forest-Tundra, reductions and northern shifts are predicted for the various classes of Taiga as well as reductions in the size of Steppe. Large increases are predicted for the Temperate Forest Steppe and Temperate Steppe vegetation zones, and for Forest Steppe and Subtaiga in the Boreal zone. Kappa statistics for judging agreement between the current climate vegetation map and the climate change predictions indicate that there is essentially no agreement between maps, thus confirming the visual impression of enormous change in the predicted vegetation of Siberia.

Acknowledgments

We are thankful to Wolfgang Cramer for generously providing the updated climate database. Helpful review comments from Nick Polikarpov, Al Solomon, and Wolfgang Cramer are appreciated.

References

Adams, R.M., C. Rosenzweig, R.M. Peart, J.T. Ritchie, B.A. McCarl, J.D. Glyer, R.B. Curry, J.W. Jones, K.J. Boote & L.H. Allen. 1990. Global climate change and US agriculture. *Nature* **346**: 219-223.

Anonymous. 1966-1970. *Reference book of climate of the USSR*. Vol. I-IV, Issues 21-24. Hydrometeoizdat, Leningrad. (in Russian).

Budyko, M.I. 1974. *Climate and life*. Academic Press, New York.

Budyko, M.I. 1986. *The evolution of the biosphere*. D. Reidel Publishing Co., Dordrecht, 423 pp.

Cohen, J. 1960. A coefficient of agreement for nominal scales. Educational and *Psychol. Measurem*. **20**: 37-46.

Conrad, V. 1947. Usual formulas of continentality and their limits of validity. *Transactions of the American Geophysical Union* **27(5)**: 663-664.

Emanuel, W.R., H.H. Shugart & M.P. Stevenson. 1985. Climatic change and the broad-scale distribution of terrestrial ecosystems complexes. *Clim. Change* **7**: 29-43.

Giorgi, F., M.R. Marinucci & G. Visconti. 1990. Use of a limited-area model nested in a general circulation model for regional climate simulation over Europe. *J. Geophys. Res*.**95(D11)**: 18413-18431.

Hansen, J., G. Russell, D. Rind, P. Stone, A. Lacis, S.Re.R. Lebedeff & L. Travis. 1983. Efficient three-dimensional global models for climate studies: models I and II. *Month. Weather Rev*. Vol. **III(4)**: 609-662.

Harrison, S.P. 1990. *An introduction to general circulation modelling experiments with raised* CO_2. Working Paper WP-90-27. International Institute for Applied Systems Analysis, Laxenburg, Austria, 52 pp.

Houghton, J.T., G.J. Jenkins & J.J. Ephraums. Eds. 1990. *Climate Change: The IPCC Scientific Assessment*. Cambridge University Press, Cambridge, UK, 365 pp.

Kolesnikov, B.P. 1965. *The zonal system of management in forestry practices. In: Problems of cultivating the Taiga zone of the USSR*. Irkutsk, Russia. (in Russian).

Krylov, A. 1984. *Life forms of forest ecosystems*. Nauka, Leningrad (in Russian), 184 pp.

Leemans, R. 1989. Possible changes in natural vegetation patterns due to a global warming. Pp 105-122 in: A. Hackl (Ed) *Der Treibhauseffekt: das Problem - Mögliche Folgen - Erforderliche Maßnahman*. Akademie für Umwelt und Energie, Laxenburg, Austria.

Leemans, R. & W.P. Cramer. 1991. *The IIASA database for mean monthly values of temperature, precipitation and cloudiness on a global terrestrial grid*. Research Report RR-91-18, International Institute of Applied Systems Analyses, Laxenburg, Austria, 62 pp.

McArdle, R.E. & W.H. Meyer. 1930. *The yield of Douglas-fir in the Pacific Northwest*. USDA Technical Bulletin No. 201. Washington, DC, 65 pp.

Manabe, S. & R.T. Wetherald. 1987. Large-scale changes in soil wetness induced by an increase in carbon dioxide. *J. Atmos. Sci.* 44: 1211-1235.

Matthews, E. 1984. *Vegetation, land-use and seasonal albedo data sets: Documentation of archived data tape.* NASA Technical Memorandum 86107, NASA, New York, 12 pp.

Matthews, E. 1985. *Atlas of archived vegetation, land-use and seasonal albedo data sets.* NASA Technical Memorandum 86199, NASA, New York, 53 pp.

Melekhov, I.S. 1980. *Forest Science.* Forest Industry Press, Moscow (in Russian), 406 pp.

Mitchell, J.F.B. 1983. The seasonal response of a general circulation model to changes in CO_2 and sea temperature. *Quart. J. R. Meteorol. Soc.* **109**: 113-152.

Monserud, R.A. 1990. *Methods for comparing global vegetation maps.* WP-90-40, International Institute of Applied Systems Analyses, Laxenburg, Austria, 31 pp.

Monserud, R.A. & R. Leemans. 1992. The comparison of global vegetation maps. *Ecol. Mod.* **62**: 275-293.

Nazimova, D.I., N.P. Polikarpov & N.M. Tchebakova. 1990. *Climatic ordination of forest vegetation zones and vertical belts as a basis for the general classification of the forest cover of Siberia.* Proc. of the International Symposium "Boreal forests: state, dynamics and anthropogenic impacts". Forest State Committee Press, Arkhangelsk-Moscow.

Orlov, M.M. 1929. *Forest Mensuration.* Silviculture and Forest Industry Press, Leningrad. (in Russian), 510 pp.

Parmuzin, Y.P. 1986. *Taiga in the USSR.* Nauka, Moscow (in Russian).

Polikarpov, N.P., N.H. Tchebakova & D. Nazimova. 1986. *Climate and Montane Forests in Southern Siberia.* Siberian Division, Academy of Science of the USSR, Novosibirsk (in Russian), 225 pp.

Prentice, I.C., W. Cramer, S.P. Harrison, R. Leemans, R.A., Monserud & A.M. Solomon. 1992. A global model based on plant physiology and dominance, soil properties and climate. *J. Biogeogr.***19**: 117-134.

Prentice, I.C., M.T. Skyes & W. Cramer. 1991. The possible dynamic response of northern forests to global warming. *Glob.Ecol. Biogeogr. Let.* **1**:129-135.

Schlesinger, M.E. & Z.C. Zhao. 1989. Seasonal climatic changes induced by doubled CO_2 as simulated by the OSU atmospheric GCM/mixed-layer ocean model. *Journal of Climate* **2**: 459-495.

Shumilova, L.B. 1962. *Botanical geography of Siberia.* Tomsk State University Press, Tomsk, (in Russian), 440 pp.

Solomon, A.M. 1986. Transient response of forests to CO2-induced climatic change: simulation modeling experiments in eastern North America. *Oecologia* **68**: 567-579.

Tchebakova, N.M., R.A. Monserud, S.V. Golovanov & R. Leemans. 1992. A global vegetation model based on the climatological approach of Budyko.. *J. Biogeogr.* **20:** 219-244.

Wetherald, R.T. & S. Manabe. 1986. An investigation of a cloud cover change in response to thermal forcing. *Clim. Change* **8**: 5-23.

Wilson, C.A. & J.F.B. Mitchell. 1987. A doubling CO_2 climate sensitivity experiment with a global climate model including a simple ocean. *J. Geophys. Res.* **92**: 13315—13343.

Yurtsev, B.A. 1986. *Relic Steppe complexes in Northeastern Asia.* Nauka, Novosibirsk, (in Russian), 166 pp.

Zubenok, L.I. 1976. *Evaporation on the continents.* Hydrometeoizdat, Leningrad, (in Russian), 264 pp.

The potential advance of pests in natural ecosystems under climate change: implications for planning and management

R W Sutherst

Centre for Tropical Pest Management, CSIRO Long Pocket Laboratories, PMB No 3 Indooroopilly, Queensland, Australia 4068

The world is too much with us, late and soon (Wordsworth)

Summary

Climate change is predicted to be evident within the next few decades, although the precise nature of the change is highly uncertain in different geographical areas. The most appropriate approach to analysing likely impacts is therefore to look for broad types of response rather than to embark on a highly detailed modelling exercise.

Pests are defined as any species which have adverse effects on the natural environment, agriculture or human well-being, or in other words any species which gets in the way of *Homo sapiens*. Such an arrogant and self centred view is central to the thinking of most policy makers today and does not bode well for the future of the world's biodiversity. In the context of natural ecosystems pests include exotic organisms such as: weeds, vertebrates, invertebrates and pathogens. Sometimes native species may also be perceived as pests.

The speed with which human-induced climate change is expected to occur and its likely continuation for centuries threatens the integrity of natural ecosystems, which have evolved over millennia. Some species will be more vulnerable than others. At the same time pests, by their responsiveness to and simultaneous tolerance of climate, high reproductive rates, short generation times, high mobility and plastic genetic composition, are well placed to exploit opportunities created by reduced competition or weakened resistance from stressed indigenous flora and fauna. By their very nature pests are likely to be some of the first species to show a response to changes in climate.

Examples are given of likely changes in the pest status of some known species of pests in protected areas. Climate change may create new species of pests. Specific threats exist from (i) the spread of pests and diseases as they follow changing climatic environments towards the poles; (ii) occupation of niches left vacant by debilitated native and exotic species; and (iii) increased incidence of permanent changes in pest distributions following extreme events such as flooding; (iv) epidemics of animal diseases as vectors spread.

Judging from experiences in pest management in agriculture, it will be technically, logistically and economically beyond the capacity of humans to manage natural ecosystems in order to preserve present levels of biodiversity in the face of climate change. Attempts to restore natural ecosytems debilitated by climate change will open up opportunities for invasive species of weeds which will out-compete desirable species that are not as well adapted to colonising disturbed areas.

Recommended responses include tightening of quarantine barriers, increased emphasis on developing the science and practice of biological control, and minimising the trend towards multi-use of protected areas where disturbance of natural ecosystems is likely. Exotic weed and marine organisms appear to pose the greatest threats to natural ecosystems under climate change.

Introduction

National parks and protected areas are established for a variety of reasons. Some protect complete ecosystems, others protect rare or attractive species, areas of natural beauty or provide human recreation. It follows that the definition of a pest in each context will differ. In the present context pests are defined as any weeds, vertebrate and invertebrate species, disease organisms and indigenous species which have adverse effects on the natural environment, native flora and fauna, or on human well-being while in the parks. It includes all species which detract from the declared purpose of the area, even native species, such as parasites which threaten the wellbeing of rare or endangered species. Valued species of mammals such as the African elephant reached damaging pest status in national parks such as the Tsavo National Park in Kenya after humans built permanent water holes. Indeed the most valued species of all - humans - are clearly the greatest pest threatening the future of protected areas. All other pest species pale into insignificance by comparison.

In the context of climate change it is also necessary to be clear about the extent of change and time scale that is relevant to any consideration of pest impacts. Much of the discussion on climate change focuses on the likely climate when the concentration of atmospheric CO_2 has doubled, in about the year 2030. Such a time frame makes it possible to develop scenarios within which to examine likely impacts in the foreseeable future. However, it is counterproductive in that it deflects the effort of the scientific community away from the much more important issue of the sustainability of the current human population and reliance on fossil fuels as an energy source. It is simply not possible to keep burning fossil fuels until they are exhausted because of the magnitude of the impact that would have on the environment. Hence all further discussion in this paper has the caveat that the impacts of pests are likely to be insignificant compared to those of humans and to continue far beyond those impacts projected for the year 2030. The scenarios for 2030 used in computer simulations for this paper are a 0.1°C increase in temperature per degree of latitude, an increase of 20% in summer rainfall and a decrease of 10% in winter rainfall from Pittock & Nix (1986). They are used only to illustrate the nature of the problems, as they will vary regionally and change as the global circulation models improve over the next decade.

Concept of climatic limits for distributions

Given the degree of uncertainty about the regional changes in climate, it is more appropriate to consider likely impacts in broad rather than specific terms. Detailed models exist for several important crop and livestock pest species, but the resolution of such models is not matched by the general circulation models (GCMs) on which climate change scenarios are based. In addition, as there are many thousands of potentially damaging pests it is not feasible to consider modelling each species in detail. Instead, the approach taken is that described by Sutherst (1991a,b) using the CLIMEX climate-matching model (Sutherst & Maywald 1985; Maywald & Sutherst 1989). CLIMEX captures the key responses of a species to climate without including detailed descriptions of the life cycle. Cammell & Knight (1991) reviewed the many aspects of pest population dynamics which would be affected by climate change.

CLIMEX can be used to either compare climates directly or to derive readily interpretable indices from the responses of a species to each of the components of climate. A weekly population growth index (GI, 0-1.0) for a species is accumulated to give an annual growth index (GI, 0-100). A series of annual *stress indices* describe the threat to the species from prolonged adverse periods of excessively cold (CS), hot (HS), dry (DS) or wet (WS) conditions. Finally the growth and stress indices are combined into an *Ecoclimatic Index* (EI), scaled from 0-100, to represent the overall favourableness of the given geographical location for the permanent survival and propagation of populations of that species. The annual number of day-degrees for development is calculated so a check can be made to see if there is enough heat accumulation during the year to complete one generation of a univoltine species or a given number of generations of a multi-voltine species.

A species' climatic requirements are inferred from information on its known geographical distribution, relative abundance and seasonal phenology. While the aim is to capture only the core features of the species' climatic requirements, CLIMEX analyses have proved very illuminating even with species that have been studied or modelled for decades. This is usually because it is very difficult to determine experimentally the precise nature of those environmental factors that limit the geographical distributions of species. Biological organisms are superb integrators of all those influences and CLIMEX attempts to simulate that process in relation to climate.

Climate change and pests

While much has been written about climate change, there has been limited research on its likely impact on pests or on the natural environment. Given climate change scenarios which are severe enough to cause changes in the structure of natural ecosystems, it is difficult to isolate the impact of pests. The climate at point A, which currently supports a given ecosystem, is likely to move several hundred kilometres towards the poles by the year 2030. It will keep moving after that period until a new equilibrium is reached after several centuries, presumably with a much impoverished biodiversity. How do we isolate the damaging role of pests during the transition of the site from the current to a new equilibrium state? Insect pests or pathogens may emerge to damage vegetation made vulnerable by the stress of a changing climatic environment (Clark

& Dallwitz, 1974; Lansberg & Wylie, 1983). If they are indigenous species, such as the bark beetles *Ips* spp in north America, and the eucalyptus beetles *Paropsis* spp. in Australia, it is likely that their activity has already contributed to the observed host distribution. In such cases it is questionable whether it is productive to make an attempt to manage the so-called *pest*. The desirable species may not be able to survive in the impending climatic environment anyway. Stands of regrowth trees with a similar age structure will be vulnerable to epidemic attacks by insects such as the spruce budworm *Choristoneura fumiferana*. On the other hand, populations of many species could proliferate should climate change upset equilibria currently maintained by natural enemies.

The situation is likely to be so complicated that it is impossible to make reliable predictions of the specific outcomes related to particular ecosystems. Indeed, unless climate change is halted, there may be no new equilibrium reached in the next millennium. A more expedient and short-term approach is to identify classes of problems which are likely to arise and pests with the potential to spread and cause increased damage to parks. Using these as test cases we can anticipate some of the types of problems which are amenable to management.

The great sensitivity of arthropod pests to climate, resulting from their high reproductive rates, short generation times, high mobility and plastic genetic composition, means that they are well placed to adapt to new environments. As such they are likely to be some of the first species to show a detectable response to changes in climate. The sensitivity of arthropod pests to climate can be expressed in different ways.

As many arthropods and plants have a single generation each year, their geographical distributions are often limited at high latitudes by a shortage of heat summation to complete their life cycle. An extended warm season suddenly releases them from that constraint and such species may become abundant in previously marginal areas and extend their ranges substantially. An example from the animal parasite world is the Far East Asian tick *Haemaphysalis longicornis*. In New Zealand the tick's range currently appears to be restricted to the North Island where there is enough heat accumulation during the summer to enable the tick to complete its life-cycle and enter the hardy nymphal overwintering stage. The tick is likely to extend its range to include most of the South Island in response to the longer warm season, given the extent of climate change that is commonly projected (Sutherst, 1990).

Species which normally increase in numbers over two or three generations a year before being stalled by the onset of winter, are able to complete an extra whole or partial generation in exceptionally warm years. This is illustrated by a tropical bovine parasite the tick *Boophilus microplus* in two years when the mean temperatures were 16.9°C and 18.5°C (Figure 1 after Sutherst, 1983). The species may cause negligible damage for several years, but then cause an outbreak in a warm year. Such phenomena are commonly observed in entomology and do not bode well for the future under global warming. Apart from the direct damage that tick feeding causes, *B. microplus* is a vector of a malaria-like protozoan parasite of cattle. In warm years the large tick numbers result in high disease transmission rates from the few infected hosts to the many younger uninfected ones, causing epidemics of disease. It is important to examine apparently benign or rare disease organisms of native animals to see if such epidemics are likely to occur in protected areas as well. Note that the above data are from years that differ by only 1.6°C in mean temperature, much less than the 2-8°C predicted for some regions by GCMs.

Many arthropods have very short life-cycles and reach their maximum population size well before the end of the growing season. In such cases it is likely that climatic warming will have less dramatic, nonlinear effects than on species with fewer generations, unless predator-prey relationships are disturbed.

Arthropod pests

The European honey bee *Apis mellifera* competes for nectar and nesting sites with many species of vertebrates and native bees in Australia (Pyke, 1990) and presumably elsewhere. The bee regulates its own breeding environment and so can colonise a wide range of climatic environments. In South America the Africanised bee has displaced the European species over much of its range except the most southerly and high altitude areas (Sheppard *et al.*, 1991). It already poses a threat to much of the USA as it can overwinter in Frankfurt, Germany (Villa *et al.*, 1991). Frankfurt has winter minimum temperatures similar to inland Oregon and Washington State in northern USA. Given global warming of 3°C, the Okanagan Valley in southern British Columbia would acquire similar winter minima. The spread of these aggressive strains could pose a threat to native fauna and flora in the Americas.

The European wasp *Vespula germanica* is less able to control the temperature in its nests in the ground and, as a result, in Europe only the queens overwinter to establish a nest in the spring (P. Spradbery, pers. comm.). A warming of temperate areas will enable entire colonies of European wasps to overwinter. In New Zealand the common wasp *Vespula vulgaris* preys on a variety of local insects and competes with them for beech honeydew in parks such as the Tongariro National Park (Fordham, 1991). Together with the European wasp they threaten native species of insects and therefore possibly insectivorous birds and plants which rely on insects for pollination (Moller & Tilley, 1989). Global warming is likely to increase wasp numbers, provided that the beech trees remain healthy. On the other hand the wasps do not appear to thrive in hot and humid environments (Spradbery & Maywald, in prep.) and their range could retract somewhat in the subtropics.

Exotic insect pests of forests, such as the gipsy moth *Lymantria dispar* in North America may cause more serious problems in higher latitudes.

Vertebrate pests

The geographical distributions of mammals and birds are less directly determined by climate than are those of poikilothermic animals. The distributions of vertebrates is more likely to be determined by changes in their habitats than by direct effects of climate. The rabbit in Australia is an example of a mammal which depends on winter growthof vegetation for its reproduction. Population growth is inhibited in hot areas with no winter rainfall (Cooke, 1977). Thus the rabbit is likely to be forced further south with global warming.

On the other hand, the distributions of reptiles are likely to respond directly to climate change. One of the most significant reptilian pests in protected areas is the giant or cane toad *Bufo marinus* which was introduced from Central America into Hawaii, some Pacific Islands and

Figure 1.
The numbers of ticks *Boophilus microplus* on cattle at Mt Tamborine in southern Queensland in two years with mean temperatures of 16.9 and 18.5°C.

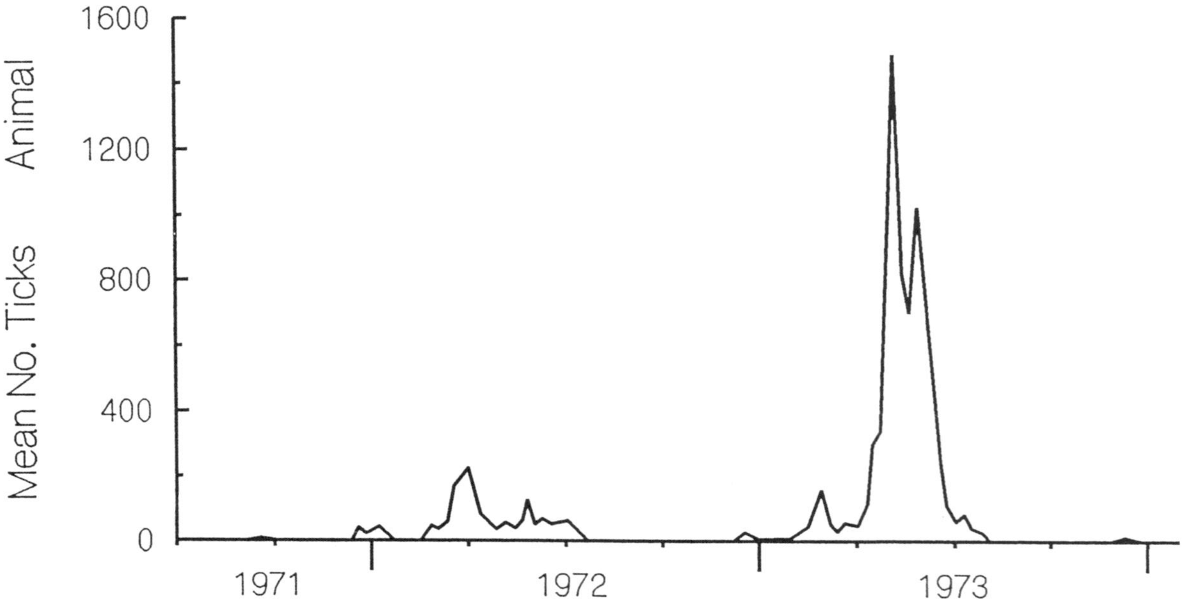

Australia in an effort to control sugar cane pests. The toad is toxic to many native predators and is also very competitive and catholic in its feeding habits. It is still spreading across northern Australia towards the Kakadu National Park with its extensive areas of wetlands which are very vulnerable to colonisation by the toad. Climate change will extend the potential of this major pest to colonise areas further south, including the Myall Lakes National Park in New South Wales (Sutherst, Floyd & Maywald, in prep.).

Fish pose one of the most significant threats to the natural environment (Moyle, 1986; Pollard, 1989). Both the aquarium and food fish trades are global and very large. There is an aggressive program of translocation of species around the world, apparently with much less quarantine surveillance or control at their destination than is customary for most other fauna or flora. The trade poses a threat to local species by importing more competitive species, predators, pathogens and diseases and even major aquatic weeds such as *Salvinia molesta*. Aggressive species such as *Oreochromis mossambicus* from southern Africa, commonly known as tilapia or mouthbrooder, have rapidly displaced local Australian species which are egg laying and surface feeders (Bluhdorn, Arthington & Mather, 1989). The lack of consideration for the environmental consequences of the trade is in urgent need of rectification, particularly because the control of fish movement and reproduction in rivers, dams and marine environments is so difficult. There is a minimum of funding for research into the environmental effects of importations, the biology of the species and their tolerance of environmental conditions. Global warming will spread the tropical problem species such as tilapia to more temperate regions. A further major source of introduction of marine organisms into new environments is the flushing of ballast tanks of bulk cargo carrying ships (Jones, 1991). Large quantities of seawater are transferred around the world in such tanks and have the potential to transfer numerous species of marine organisms to new environments. Warming of the oceans will affect the success of such fish, invertebrates and plants in the same way as it will terrestrial organisms. There will be a tendency for all species to move towards temperate zones.

Animal parasites

The balance between animals and their parasites is easily disturbed by changes in the grazing patterns of herbivores. Large ungulates in Africa remain largely unaffected by the tick *Rhipicephalus appendiculatus* when allowed to move freely over their extensive home ranges. However, when that movement is restricted tick numbers can build up to very damaging levels as observed in the Vernon Crookes, Natal Lion Park and Umgeni Valley Nature Reserves in South Africa (Lewis, 1981) and the Kyle Recreational Park in Zimbabwe (Lightfoot & Norval, 1981). Most tick-related problems of wild animals in Africa arise from *R. appendiculatus* which does not tolerate high temperatures or dryness. Such problems are likely to occur at higher altitudes and latitudes following global warming.

Poor nutrition is one of the stresses which reduce animal host resistance to parasite infestations (Sutherst *et al.*, 1983). Moose in the southern part of their range in Canada are frequently heavily infested with the tick *Dermacentor albipictus*, often in association with poor host nutrition. Infested moose remove much of their hair during grooming giving rise to the phenomenon known as 'ghost moose'. In the Elk Island National Park in Alberta, Canada, 88% of moose suffered hair loss, which averaged 20-40% per animal (Samuel & Welch, 1991). In Long Lake Provincial Park in Alberta 100% of 20 moose examined had tick-induced hair loss. The tick is distributed as far as the 60°N latitude line (Anderson & Lankester, 1974) and would move even further north after global warming. Nutritional stress is likely to accompany dislocation of moose grazing patterns in protected areas with limited scope for migration. This will exacerbate the tick problem.

On the other hand the avian tick *Ixodes uriae* reaches very high densities in nesting areas in Macquarie Island near Antarctica. What will happen when the environment becomes too warm for the ticks which require temperatures of 0-5°C? The ticks could be lost together with any regulating effect they may have on the bird populations.

Weeds

Invasion of disturbed areas by exotic weeds is a major concern of most national park authorities (Usher, 1986; Humphries, Groves & Mitchell 1991). Many exotic ornamental plants with the potential to become weeds are moved between countries with minimum consideration of environmental impact. Global warming is likely to exacerbate the weed problem by debilitating native vegetation and so reducing competition. For example after a prolonged drought in southern Queensland in 1991-2 a severe dieback occurred of native forest trees, opening up the canopy and stimulating the growth of exotic species such as *Lantana camara* and *Protoasparagus* (P. Grimshaw, pers comm,).

In northern Australia the Kakadu National Park is currently under threat from a number of weeds. One of the most important is the floating fern, *Salvinia molesta* from Brazil. It has invaded some of the park's waterways and threatens others. Studies to manage the weed, including the use of biological control, are currently taking place.

Meanwhile, the terrestrial 'giant sensitive plant' *Mimosa pigra*, from tropical America, is spreading rapidly in Australia, after escaping into the Adelaide River delta 60 years after its original introduction into the area. This follows habitat modification by feral animals and extensive flooding in the mid-1970s (Lonsdale, Miller & Forno, 1989). The problem provides an example of a change in pest status of a plant following the type of extreme climatic event that is predicted to become more frequent in future (Wigley, 1985). *M. pigra* is currently the subject of an eradication campaign in the neighbourhood of Kakadu National Park. The plant variety in Australia occurs naturally around river and coastal systems from Rio de Janeiro in Brazil north to Veracruz in Mexico. It has been introduced into Florida where it threatens the Everglades and South East Asia, where the Doi Suthep and Doi Inthanon National Parks in Thailand are severely affected (W. Forno, pers. comm.).

Based on the relative abundance of *M. pigra* over its known range, it is possible to estimate its present and future potential as an invader of tropical and subtropical areas around the world. The CLIMEX model was used to define the present climatic conditions which favour the growth and persistence of *M. pigra*, as inferred from its geographical distribution (Figure 2a). Given these assumptions, it has the potential to spread greatly around the Old World. The potential for spread of the shrub under the influence of global warming can be inferred as shown in (Figure 2b). Clearly, the species poses a major threat to a large part of the region's natural ecosystems.

In the USA two vines, kudzu *Pueraria lobata* and Japanese honeysuckle *Lonicera japonica* have been investigated in relation to climate change (Sasek & Strain, 1990). The former species causes damage in forest plantations and in any disturbed, unused land. The honeysuckle is invasive in parks and gardens. Both have become established over much of the south east of the USA, with the honeysuckle causing its most severe problems in the area shown in Figure 3a. The CLIMEX model was used to define the climatic requirements and preferences of the vine (Fig 3a) and a projection was made of the likely spread of the species given a 3°C rise in temperatures (Figure 3b). The weed would migrate up to the Great Lakes and increase in severity in the north-eastern USA, affecting Parks such as the Blue Ridge Mountain Park in Virginia. The results are similar to those of Sasek & Strain (1990) based on changes in isoclines of rainfall and temperature.

Pathogens

Warm and humid conditions favour the growth of many plant pathogens. A particularly severe pathogen which poses a major threat to native plants and forests in many parts of the world is *Phytophthora cinnamomi*. Increased incidence of high rainfall events with temperatures above 20°C will favour outbreaks of this pathogen, as growth is maximised at high soil moisture levels with temperatures between 25 and 30°C, following prolonged dry spells. Given a 3°C rise in temperature with the associated rise in evaporation rates, and 20% increase in summer rainfall in the subtropics, the incidence of extremely dry or wet weather events will increase disproportionately (Figure 4) and could be expected to lead to more outbreaks of the disease. As the

Figure 2.
The potential global distribution of *Mimosa pigra* as indicated by the value of the CLIMEX ecoclimatic index (EI): (a) at present and (b) in the year 2030. The values of EI are proportional to the size of the circles and zero values are indicated by crosses.

(a)

(b)

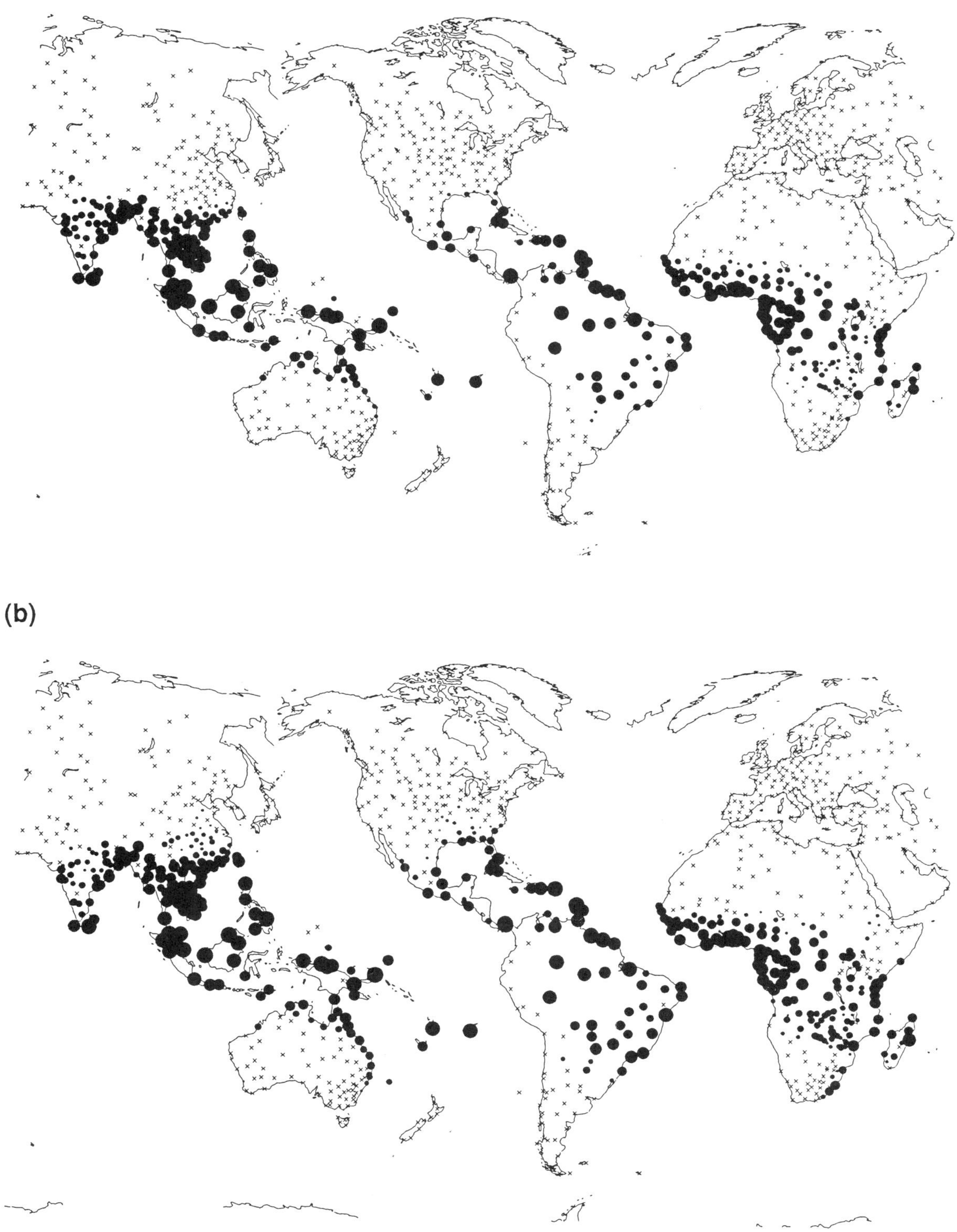

EI : Mimosa pigra

Figure 3.
The potential distribution of Japanese honeysuckle in North America as indicated by the value of the CLIMEX ecoclimatic index (EI) (a) at present and (b) after climate change in the year 2030. The values of EI are proportional to the size of the circles and zero values are indicated by crosses.

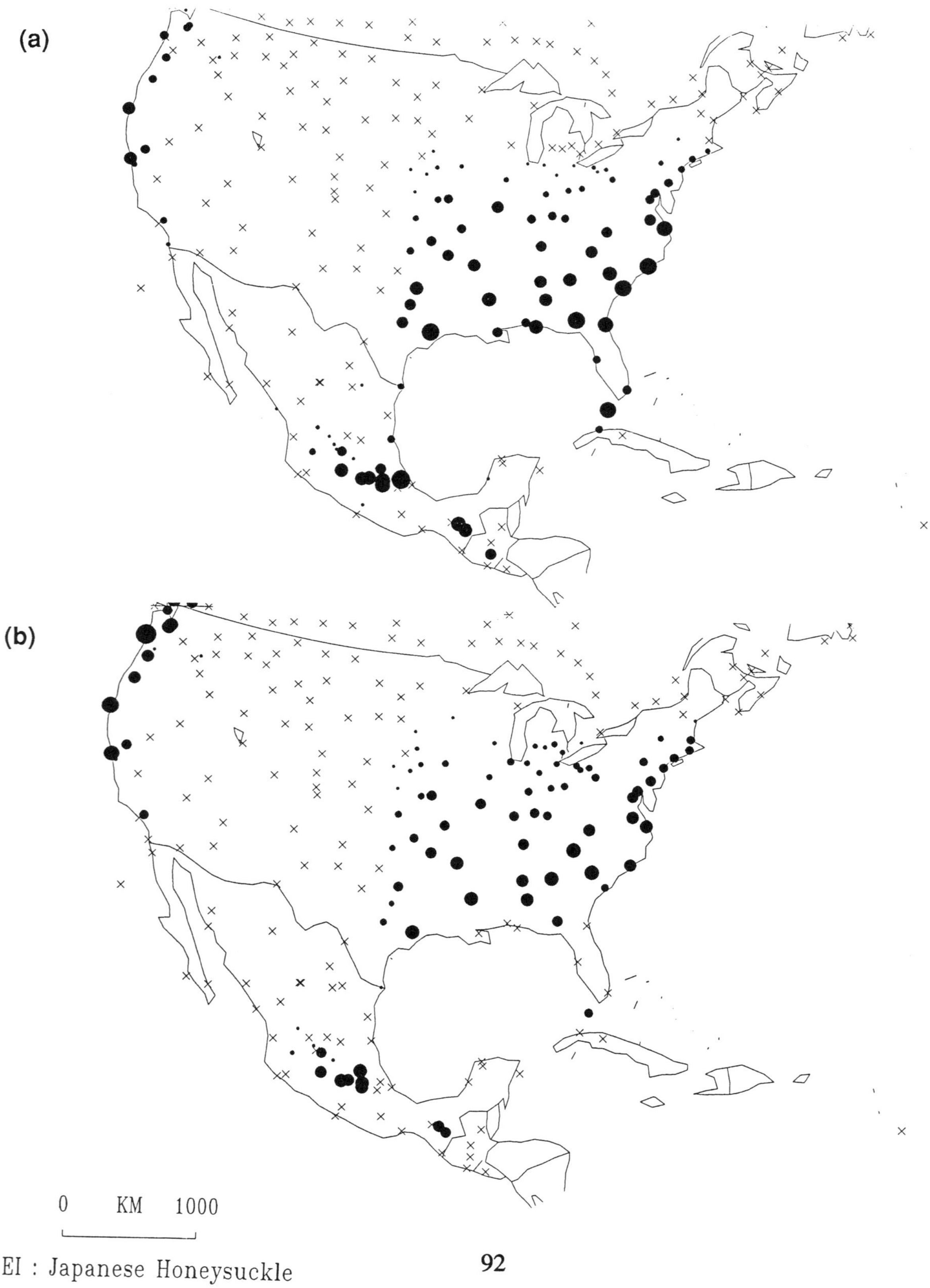

fungus is moved mainly by human activity and attacks indigenous eucalypt trees in Australia it is a serious concern to managers of native forests in South-west Australia (Shearer & Tippett, 1989), Victoria (Marks & Smith, 1991) and the other States.

Human nuisance pests

Many protected areas are established essentially for human recreational use, rather than to protect native plants or animals. In these areas conservation measures have resulted in increased numbers of small and large mammals, which are reservoirs of several debilitating human conditions such as Lyme disease and tick-borne typhus and encephalitis. In the USA the lone star tick *Amblyomma americanum* causes severe problems in recreational areas such as the Ozark areas in Oklahoma (Hair & Howell, 1970) and Land Between the Lakes, Kentucky-Tennessee (Bloemer *et al.*, 1990). Using CLIMEX, it can be shown that global warming will result in the focus of those problems being transferred north, as shown above for the Japanese honeysuckle. Similar tick-borne diseases affect users of protected areas in many parts of Europe where the tick *Ixodes ricinus* is widespread.

Some species of ticks, such as *Ixodes holocyclus* in Australia, cause severe allergies and even lethal paralysis to humans visiting national parks on the east and southern coastal strip of the continent. Global warming will increase those problems in the more densely populated southern parts of the continent.

Figure 4.
Frequency distributions of extreme high rainfall months at Lawes (Southern Queensland) compared with the expected distribution given a 20% increase in rainfall.

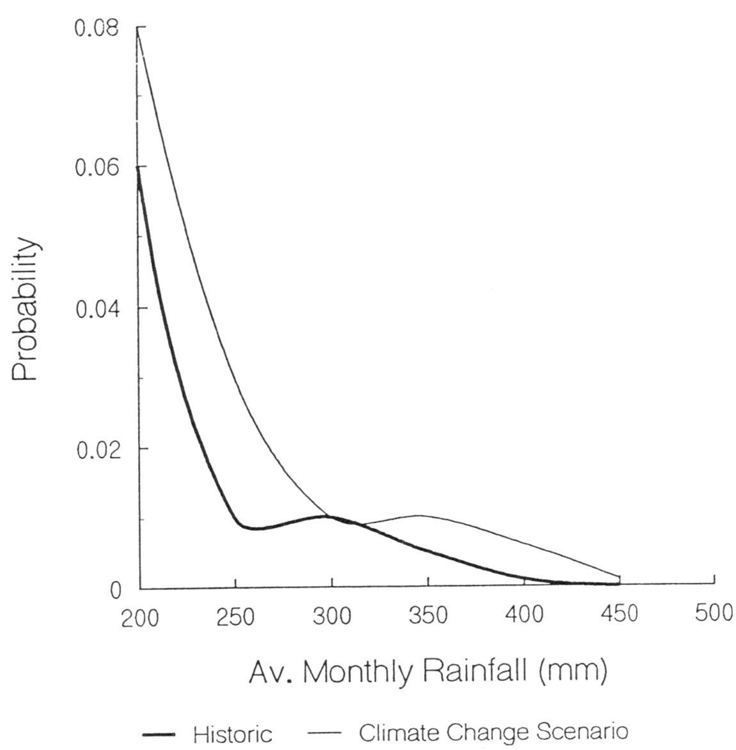

In many humid vegetation types around the world leeches are a nuisance to visitors and their incidence is likely to change, but in ways that will vary depending on whether the area concerned receives more or less rain. With the increased temperatures more rain will be needed simply to offset the increased evaporation and so maintain current levels of soil moisture.

Planning and management

Had we but world enough, and time (Andrew Marvell)

Threats exist from (i) spread of pests to follow existing climatic environments as they are pushed towards the poles; (ii) vacant niches left by less competitive or dying native species; (iii) changes in pest status following extreme events; and (iv) epidemics of vector-borne diseases. A number of responses are recommended below.

As mentioned above, the precise nature of the processes by which climatic or other factors limit geographical distributions have rarely been defined for even the most thoroughly studied species. An analysis of the geographical distributions of pests and the limiting factors involved should lead to economies in the production of databases as many species have similar distributions and limiting factors. Comparative studies will help to define the nature of limiting factors more precisely, and so facilitate future application to other species.

Planning requires much more information on the potential of pests/weeds to invade and colonise protected areas in response to climate change. We need to define the global areas at risk, as shown for the Japanese honeysuckle in Figure 5. Production of global databases, which contain descriptions of areas at risk from key pests, and the nature of limiting effects, such as those produced by CLIMEX, would be an invaluable resource in a world with or without a changing climate. The use of Geographic Information Systems and expert systems to integrate variables on a spatial scale will enhance the value of that information to policy makers (Sutherst *et al.*, 1991). Such computer-based techniques are the basis for powerful decision-support systems in agricultural pest management.

To complement these efforts we need to define those characteristics of each exotic pest group that enable them to be more competitive than indigenous species.

There will be a need for increased reliance on expertise of scientists with tropical experience as the climate warms. Steps need to be taken to redress the present imbalance between temperate and tropical ecological research and training.

Disturbance from multipurpose use of protected areas carries high risks of providing areas opportunities invasion by pests such as *Phytophthora*. Every effort should be made to keep those areas, which are set aside to protect natural ecosystems, in pristine condition.

One of the most important barriers to colonisation by exotic fauna is through quarantine measures, which should be strengthened by placing more emphasis on environmental as opposed to economic risks. Buffer zones are needed around sensitive local habitats such as caves and lakes to prevent the entry of exotic species.

Figure 5.
The potential global distribution of the Japanese Honeysuckle as indicated by the value of the CLIMEX ecoclimatic index (EI). The values of EI are proportional to the size of the circles and zero values are indicated by crosses.

EI : Japanese Honeysuckle

Accepting the inevitability of the continuing entry of exotic species into protected areas, there is a need to develop the science and practice of biological control. Biological control is likely to be the only sustainable method of pest control when pests have become well established over large protected areas. It will be the best way ofreducing the competitive advantage enjoyed by many exotic species which are usually introduced into new areas without their associated herbivores, parasites or predators.

Conclusions

The major pests of protected areas are weeds and large mammalian herbivores and predators. Disturbance of natural vegetation by climate change is likely to provide the opportunity for weeds to dramatically increase their ability to invade natural ecosystems. Fish and ornamental plants are equally important because of the apparent difference in quarantine standards applied to these groups compared with other taxa and the ease with which shipping transfers species of marine organisms around the world in ballast tanks. Exotic arthropod pests are less often cited as problems in protected areas, probably because they usually restrict their feeding to exotic plants.

While the risk to natural ecosystems from the spread of pests as a result of climate change is high, it eclipsed by the threat from direct human impacts in the face of human population growth and industialisation. Global warming is likely to progress as long as humans continue to increase in numbers and maintain their consumption of fossil fuels. We need to acknowledge that humans are the ultimate pests that need to be managed if protected areas are to survive in something like their present form. Human population growth is probably the largest single threat to planet earth, and yet the World Bank for example spends only 0.4% of its budget on birth control in the face of a huge unsatisfied demand for control of unwanted fertility (Short, 1991). Similarly, Lovins (1989) states that full use of best available energy-saving technology could reduce electricity, oil and gas consumption in the USA by 75% at less cost than finding new energy sources.

The overwhelming management problems associated with preserving natural ecosystems in the face of human population growth and industrialisation puts the onus back on society to address the source of the problem rather than on the symptoms. Biologists must be at the forefront of community efforts to counter the Greenhouse Effect and other global impacts of human population growth, poverty and resource exploitation. If we concentrate our efforts on adapting to climate change *per se* we are liable to miss the real problem. Climate change and depletion of the ozone layer are merely the most obvious symptoms of environmental degradation. We need to address the fundamental cause, not the symptoms, if the world is to remain a fit place for living organisms, including humanity. However, we also need to know how best to live with the symptoms, which are now unavoidable to some extent, otherwise the other pests will inherit the earth even sooner than we expect.

Acknowledgments

My thanks to the many people who so willingly provided me with information, including Rob Floyd, Wendy Forno, Greg Gowing, Paul Grimshaw, Mic Julien, Hank Jenkins and Rolley McKay. Gunter Maywald assisted with the preparation of the illustrations.

References

Anderson, R.C. & Lankester, M.W. 1974. Infectious and parasitic diseases and arthropod pests of moose in North America. *Naturaliste Canadien* **101**:23-50.

Bloemer, S.R., G.A. Mount, T.A. Morris, R.H. Zimmerman, D.R. Barnard & E.L. Snoddy. 1990. Management of lone star ticks (Acari: Ixodidae) in recreational areas with acaricide applications, vegetative management, and exclusion of white-tailed deer. *J. Med. Entomol.* **27**: 543-550.

Bluhdorn, J., A.H. Arthington & P.B. Mather. 1989. The introducedcichlid, *Oreochromis mossambicus*, in Australia: a review of distribution, population genetics, ecology, management issues and research priorities. In Pollard, D.A. (Ed) *Introduced and translocated fishes and their ecological effects*. Bur. Rural Resources Proc. No.8. Australian Government Printer, Canberra.

Cammell, M.E. & J.D. Knight. 1991. Effects of climatic change on population dynamics of crop pests. *Adv. Ecol. Res.* **22**.

Clark, L.R. & M.J. Dallwitz. 1974. On the relative abundance of some Australian Psyllidae that coexist on *Eucalyptus blakelyi*. *Aust. J. Zool.* **22**:387-415.

Coke, B.D. 1977. Factors limiting the distribution of the wild rabbit in Australia. Pp 113-120 in: D. Anderson (Ed) *Exotic species in Australia - their establishment and success.* Ecological Soc. Australia.

Fordham, R.A. 1991. Vespulid wasps at the upper forest margin in Tongariro National Park. *New Zealand J. Zool.* **18**:151-153.

Hair J.A. & D.E. Howell. 1970. *Lone star ticks: their biology and control in Ozark recreational areas.* Okla. Agric. Exp. Stn. Bull. B-679, 47pp.

Humphries, S.E., R.H. Groves & D.S. Mitchell. 1991. *Plant invasions of Australian ecosystems.* Unpubl. Report to Australian National Parks and Wildlife Service. Endangered species program Project No. 58.

Jones, M.M. 1991. *Marine organisms transported in ballast water: a review of the Australian scientific position.* Bur. Rural Resources Bull. No. 11. Australian Government Printer, Canberra.

Landsberg, J. & F.R. Wylie. 1983. Water stress, leaf nutrients and defoliation: a model of dieback of rural eucalypts. *Aust. J. Ecol.* **8**:27-41.

Lewis, A.R. 1981. The Pathology of *Rhipicephalus appendiculatus* infestation of Eland *Taurotragus oryx.* In: G.B. Whitehead and J.D. Gibson (Eds) *Tick Biology and Control.* Rhodes University, Grahamstown.

Lightfoot, C.J. & R.A. Norval. 1981. Tick problems in wildlife in Zimbabwe. 1. The effects of tick parasitism on wild ungulates. *S. Afr. J. Wildl. Res.* **11**:41-45.

Lonsdale, W.M., I.L. Miller & I.W. Forno. 1989. The biology of Australian weeds 20. *Mimosa pigra* L. *Plant Prot. Quart.* **4**:119-131.

Lovins, A.B. 1989. *Energy, people, and industrialization. Unpubl. paper commissoned for Conference on Human demography and natural resources.* Hoover Institution, Stanford. 26pp.

Moyle, P.B. 1986. Fish introduction into North America: Patterns and ecolgical impact. Pp 27-43 in: Mooney, H.A. & Drake, J.A. (Eds) *Ecology of Biological Invasions of North America and Hawaii.* Springer-Verlag, New York.

Pollard, D.A. Ed. 1989. *Introduced and translocated fishes and their ecological effects.* Bur. Rural Resources Proc. No.8. Australian Government Printer, Canberra.

Marks, G.C. & I.W. Smith. 1991. The cinnamon fungus in Victorian forests. *Victorian Lands & Forests Bull.* **31**: 33pp.

Maywald, G.F. & R.W. Sutherst. 1989. CLIMEX, a computer program for comparing climates in ecology: recent developments. Pp 134-140 in: *Proc. SSA IMACS 1989 Biennial Conference on Modelling and Simulation*, Canberra. pp.134-140.

Moller, H. & J.A.V. Tilley. 1989. Beech honeydew: seasonal variation and use by wasps, honey bees and other insects. *New Zealand J. Zool.* **16**:289-302.

Pyke, G. 1990. Apiarists versus scientists: a bittersweet case. *Australian Natural History* **23**: 386-392.

Pittock A.B. & H.A. Nix. 1986. The effect of changing climate on Australian biomass production - a preliminary study. *Clim. Change* **8**:243-255.

Samuel, W.M. & D.A. Welch. 1991. Winter ticks on moose and other ungulates: factors influencing their population size. *Alces* **27** (in press)

Sasek, T.W. & B.R. Strain. 1990. Implications of atmospheric CO_2 enrichment and climatic change for the geographical distribution of two introduced vines in the U.S.A. *Clim. Change* **16**:31-51.

Shearer,B.L. & J.T. Tippett. 1989. *Jarrah dieback: The dynamics and management of Phytophthora cinnamomi in the Jarrah (Eucalyptus marginata) forest of south-western Australia.* Western Australia Dept Conservation and Land Management Bull 3: 76pp.

Sheppard, W.S., T.E. Rinderer, J.A. Mazzoli, J.A. Stelzer & H. Shimanuki. 1991. Gene flow between African- and European-derived honey bee populations in Argentina. *Nature* **349**:782-784.

Short, R. 1991. Human population growth. *Austr. Biol.* **4**: 215-219.

Sutherst, R.W. 1983. Variation in the numbers of the cattle tick, Boophilus microplus (Canestrini), in a moist habitat made marginal by low temperatures. *J. Aust. Entomol. Soc.* **22**: 1-5.

Sutherst, R.W. 1990. Impact of climate change on pests and disease in Australasia. *Search* **21**: 230-232.

Sutherst, R.W. 1991a. Predicting the survival of immigrant insect pests in new environments. *Crop Protection* **10**: 331-333.

Sutherst, R.W. 1991b. Pest risk analysis and the greenhouse effect. *Rev. Agric. Entomol.* **79**:1177-1187.

Sutherst, R.W., J.D. Kerr, G.F. Maywald & D.A. Stegeman. 1983. The effect of season and nutrition on the resistance of cattle to the host, Boophilus microplus. *Aust. J. Agric. Res.* **34**: 329-39.

Sutherst, R.W. & G.F. Maywald. 1985. A computerised system for matching climates in ecology. *Agricult. Ecosyst. Environm.* **13**:281-299.

Sutherst, R.W., G.F. Maywald & W. Bottomley. 1991. From CLIMEX to PESKY, a generic expert system for pest risk assessment. *EPPO Bull.* **21**: 595-608.

Usher, M.B. 1986. Invasibility and wildlife conservation: invasive species on nature reserves. *Phil. Trans. R. Soc. Lond. B* **314**:695-710.

Villa, J.D., N. Koeniger & T.E. Rinderer. 1991. Overwintering of Africanised, European and hybrid honey bees in Germany. *Environ. Entomol.* **20**:39-43.

Wigley, T.M.L. 1985. Impact of extreme events. *Nature* **316**:106-107.